# FUNDAMENTALS OF AIRCRAFT MATERIAL FACTORS

## SECOND EDITION

## by Charles E. Dole

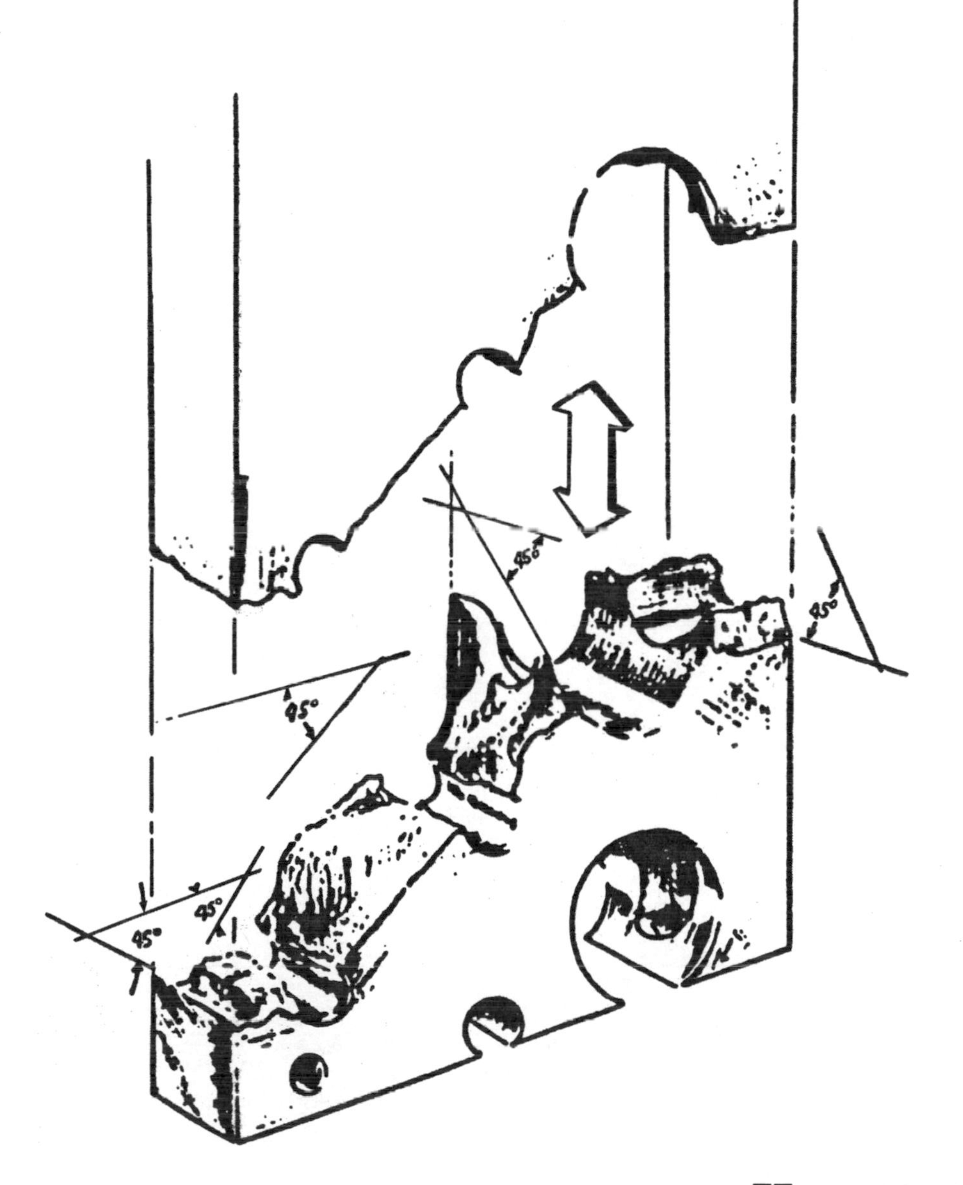


JEPPESEN®

Jeppesen Sanderson, Inc.

Published in the United States of America
Jeppesen Sanderson, Inc.
55 Inverness Drive East, Englewood, CO 80112-5498
www.jeppesen.com



ISBN-13: 978-0-89100-340-3
ISBN-10: 0-89100-340-1

Jeppesen Sanderson, Inc.
55 Inverness Dr. East
Englewood, CO 80112-5498
Web Site: www.jeppesen.com
Email: Captain@jeppesen.com

Printed in the United States of America

JS312646-001

# PREFACE

"AVIATION SAFETY IS NO ACCIDENT" - this banal treatment of an important subject may offend the reader's intelligence, but it does make a point. Accident prevention is the prime goal of aviation safety.

To prevent accidents we must first find out what causes them. As the causes are varied, the factors involved cover many fields. Some of these are listed in alphabetical order.

Accident Investigation
Aerodynamics
Aircraft Design
Aircraft Manufacture
Aircraft Maintenance
Communications
Human Engineering
Legal Aspects
Material Factors
Meterology
Photography
Physiology
Psychology
Safety Program Management
Systems Safety Engineering

Accident investigation is a valuable tool in the prevention of future accidents. We learn (or should learn) from our mistakes. To make a meaningful analysis of an aircraft mishap, the accident investigator should have a sound knowledge of the engineering principles involved in the design, construction, maintenance, and operation of aircraft. The approach to teaching the basics of these technological principles to USAF Flying Safety Officer Course and the USAF Mishap Investigation Course students is two-fold: (1) a study of Aerodynamics and (2) a study of Material Factors.

Aerodynamics is covered in *FLIGHT THEORY FOR PILOTS*, another text by the author. This book covers the second of these subjects.

The subjects covered in the book include:

Design Specifications for Aircraft
Loads on Aircraft
V-G Diagram
Metallic Composition
Elementary Stress Analysis
Stresses from Bending and Torsion Loads
Failures in Riveted Joints and Pressure Vessels
Strain and Stress-Strain Diagram
Compressive Failures
Aircraft Metals and their Heat Treatment
Composite Materials
Stress Concentrations
Fatigue of Aircraft Materials
Corrosion
Non-Destructive Inspection

The book is written as a text and/or guide for aircraft accident investigators and members of accident boards. It covers the basics of material factors involved in mishaps primarily, but not exclusively, of aircraft. Many of the principles covered apply to all structural design, not only to aircraft. It is not intended to be a complete coverage of the subjects presented. For more detail, the reader should consult texts such as those listed in the References section at the end of the book.

I am grateful for the assistance of my fellow faculty members, especially Mr. Gay Jones and Mr. Walter Schob who helped with suggestions and criticism of the manuscript.

Charles E.Dole, EdD
Field Associate Professor
University of Southern California
October, 1986

# PREFACE TO SECOND EDITION

As discussed in the Preface above, this book was originally written as a textbook for students of military aircraft accident investigation. Subsequently it was discovered that there was a demand for it in the commercial aviation field, as well. Thus, more references to Federal Regulations have been included in this edition. An appendix of thirty pages showing more examples of various metal failures has been added. Improved line drawings are also used in this edition.

The third printing includes use of a new scientific word processor, a revised chapter on Composite Materials and other revisions.

The fourth printing uses a laser printer.

Charles E. Dole, EdD
July, 1989

# CONTENTS

## CHAPTER SIX
## STRESSES FROM TORSION AND BENDING LOADS



## CHAPTER SEVEN
## RIVETED JOINTS AND PRESSURE VESSELS



## CHAPTER EIGHT
## STRAIN AND THE STRESS-STRAIN DIAGRAM



## CHAPTER NINE
## COMPRESSION FAILURES



## CHAPTER TEN
## AIRCRAFT METALS AND THEIR HEAT TREATMENT

## CHAPTER ELEVEN
## COMPOSITE MATERIALS



## CHAPTER TWELVE
## STRESS CONCENTRATIONS



## CHAPTER THIRTEEN
## FATIGUE OF AIRCRAFT MATERIALS



## CHAPTER FOURTEEN
## CORROSION



## CHAPTER FIFTEEN
## NONDESTRUCTIVE INSPECTION

# CHAPTER ONE

## DESIGN SPECIFICATIONS FOR AIRCRAFT

Military Standard MIL-~~STD-1530A(11)~~ establishes the AIRCRAFT STRUCTURAL INTEGRITY PROGRAM, (ASIP), AIRPLANE REQUIREMENTS for all departments and agencies of the Department of defense.

It specifies acceptance standards to be met by contractors conducting the development of airframes for (1) manned power driven aircraft having fixed or adjustable fixed wings and (2) to those portions of manned helicopter and V/STOL aircraft that have similar structural characteristics.

The Standard also applies to government personnel in managing the development, production, and operational support of airplane systems throughout their life cycles.

The contractor who manufactured the aircraft was required to conduct many categories of tests to assure that the airplane met the specifications including "structural" flight tests.

These tests verified the structural limit loads to the extremes of the design envelope.

Loads beyond the positive and negative limit loads were imposed in static ground tests to verify the design ultimate strength capabilities of the airframe.

Besides the Military Standard for the entire ASIP, there are many Military Specifications that cover detailed requirements.

These are found in the MIL-A-88xx series. Pertinent extracts will be covered in the text.

Airworthiness standards of civilian aircraft are covered in the Code of Federal Regulations, Title 14

Part 23 (Normal, Utility, Acrobatic Category Airplanes),
Part 25 (Transport Category Airplanes),
Part 27 (Normal Category Rotorcraft),
Part 29 (Transport Category Rotorcraft).

Personnel involved in flying (aviation) safety programs such as Flying (or Aviation) Safety Officers and aircraft accident investigators, should have a knowledge of the various specifications/regulations.

Such knowledge will help them understand the limitations of aircraft and the importance of not exceeding them.

## *Design Specifications for Aircraft*

Although there is no official distinction between the following three areas of structural design, here they are separated for clarity.

### 1. Static Strength

A static load can be thought of as a single gradually applied load. Repeated loads are considered under the area of Service Life and will be discussed later.

Mil-A-8861(ASG) discusses flight loads for military aircraft and Parts 23, 25, 27 and 29 of Title 14 cover civilian aircraft flight loads.

*Load Factor* ("G" or "n") is the ratio of the load (L) on the aircraft to the weight (W) of the aircraft.

$$G = \frac{L}{W} \tag{1.1}$$

*Load* is the total force acting on the airplane.

$$L = G \times W \tag{1.2}$$

*Limit Load Factor* (LLF) is: "the maximum load factor authorized for operations."

*Limit Load* (LL) is:

$$LL = LLF \times W \tag{1.3}$$

LL is constant for all weights above design gross weight. The limit load factor is reduced if gross weight is increased. But, the LLF cannot be increased if the gross weight is decreased below the design gross weight. Engine mounts and other structural members are designed for the nominal LLF.

*Ultimate Load* (UL) is: "Ultimate loads are obtained by multiplying the limit loads by the ultimate factor of safety. Failure shall not occur at the ultimate load.The ultimate factor of safety shall be 1.50."

$$ULF = 1.50 \times LLF \tag{1.4}$$

The manufacturer of the aircraft was required to conduct many categories of tests to assure that the airplane met the specifications/regulations including "structural" flight tests. These tests verified the structural limit loads to the extremes of the design envelope.

Loads beyond the positive and negative limit loads were imposed in static ground tests to verify the design ultimate strength capabilities of the airframe.

It should be kept in mind that the tests were conducted upon new airplanes also that these airplanes undoubtedly were constructed with more care than the average production model. No corrosion or fatigue problems existed for these test airplanes. Older fleet airplanes may have reduced capability to withstand high load factors. Treat the limits with caution.

From Table I of Mil-A-8861 for symmetrical flight LLF at basic flight design gross weight show:

| | | |
|---|---|---|
| Attack aircraft | +8.67, | -3.00 |
| Fighters and Fighter Trainers | +7.33, | -3.00 |
| Bombers (Light) | +3.67, | -1.67 |
| Bombers (heavy) | +2.00, | 0 |
| Cargo (assault) | +3.00, | -1.00 |
| Cargo (transport) | +2.50, | 0 |

Examples of Load Factor Limitations of T-38A

| Gross Wt. | Symmetrical LLF | | Unsymmetrical LLF | |
|---|---|---|---|---|
| 12,500 lb. | -2.3 | +5.6 | 0 | +4.0 |
| 11,000 lb. | -2.6 | +6.4 | 0 | +4.6 |
| 9,600 lb. | -3.0 | +7.33 | 0 | +5.22 |

Limit Load = LLF x W = 5.6 x 12,500 = 70,000 lb

It is seen that the above load factors are different for symmetrical and unsymmetrical flight. During unsymmetrical maneuvers, such as a rolling pull-out, the wing that is rising will have more lift on it, and thus a greater load factor, than the downgoing wing.

This is shown in Figure 1.1. The accelerometer is located on the centerline of the aircraft and will measure the average load factor. In Figure 1.1 the pilot reads 5 G's, but the rising wing will be subjected to 7 G's.

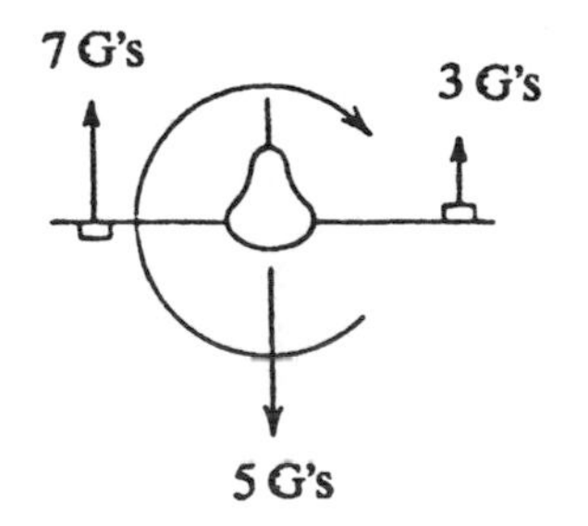


*Figure 1.1 Unsymmetrical loading.*

Sometimes (e.g., T-38A) the downgoing wing is limited by torsion in that the up aileron travels farther than the down aileron and thus applies more torsion to the wing structure.

Typical Limit Load Factors for civilian aircraft are: G - LOADS

| | | |
|---|---|---|
| Normal Category | +3.8, | -1.5 |
| Utility | +4.4, | -1.8 |
| Acrobatic | +6.0, | -3.0 |
| Transport | +2.5 to +3.8, | -1.0 |

## * 2. Rigidity Requirements

*Rigidity* is the resistance to deflection under load.

Besides adequate static strength the airplane must have enough rigidity to resist excessive deflections of the airframe due to static and dynamic loads. The general requirement, as stated in Mil-A-8870, is: "Construction, materials, and design shall be such that there will be no flutter, buzz, or other dynamic instabilities, divergence, or excessive vibration of the structural components of the airplane throughout the design range of altitudes, maneuvers, thermal conditions where losses in rigidity may occur, etc."

## Aeroelastic Problems

Most of the problems that have occurred in the past, due to a lack of rigidity, can be grouped under the general title of *Aeroelastic Problems.*

A. *Aileron Reversal* was a major problem in the first large swept wing aircraft (e.g., B-47). The designers had not yet overcome the structural problems of providing the necessary bending rigidity into a large swept wing.

A symmetrical wing section develops an upward aerodynamic force (AF) at the aileron hinge point when the aileron is deflected downward. This is shown in Figure 1.2. The AF acts behind the elastic axis and so produces a nose down pitching moment. This moment depends on velocity squared, thus is greatest at high speeds.

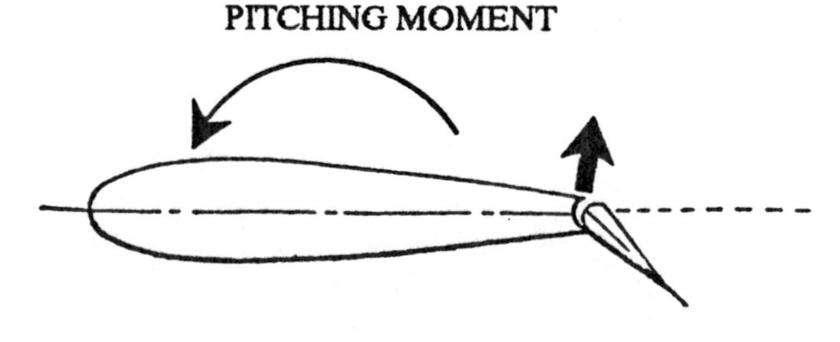


***Figure 1.2 Deflected aileron produces a pitching moment.***

If the wing lacks sufficient rigidity to resist the pitching moment it is called an "elastic wing" and will be twisted nose downward as shown in Figure 1.3. It will then be operating at a lower angle of attack and thus suffer a reduction in lift. At some high airspeed the moment and resulting twist may be great enough to cause the wing to lose lift when the aileron is deflected downward.

This is called *aileron reversal.*

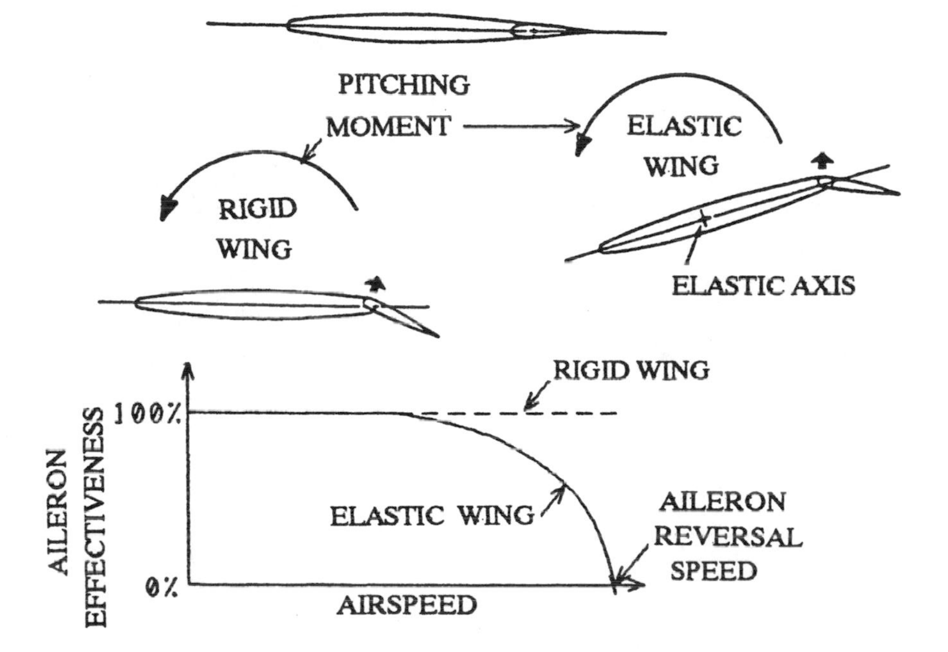


***Figure 1.3 Aileron Reversal.***

*B. Divergence* is another aeroelastic problem arising from lack of wing twisting rigidity. Here the cause of the twisting moment is wind gusts. If an up-gust strikes the wing an increase in angle of attack results. Lift, is increased. It effectively acts at the aerodynamic center (AC), (located at approximately 25 percent of the chord behind the leading edge). The wing twists nose upward about its elastic axis (located some distance behind the AC). This increases the angle of attack and lift even more and the self-perpetuation continues. At some combination of high airspeed, high up-gust and low rigidity, the wing tips may depart (diverge) from the aircraft. This is shown in Figure 1.4. Wing divergence has been the major reason that swept forward wings have not been developed for jet aircraft until recently. Forward swept wings accentuate the divergence problem.

New techniques (discussed in Chapter 11) using composite materials have produced wing structures that resist twisting moments but still permit upward flexing under gust loads.

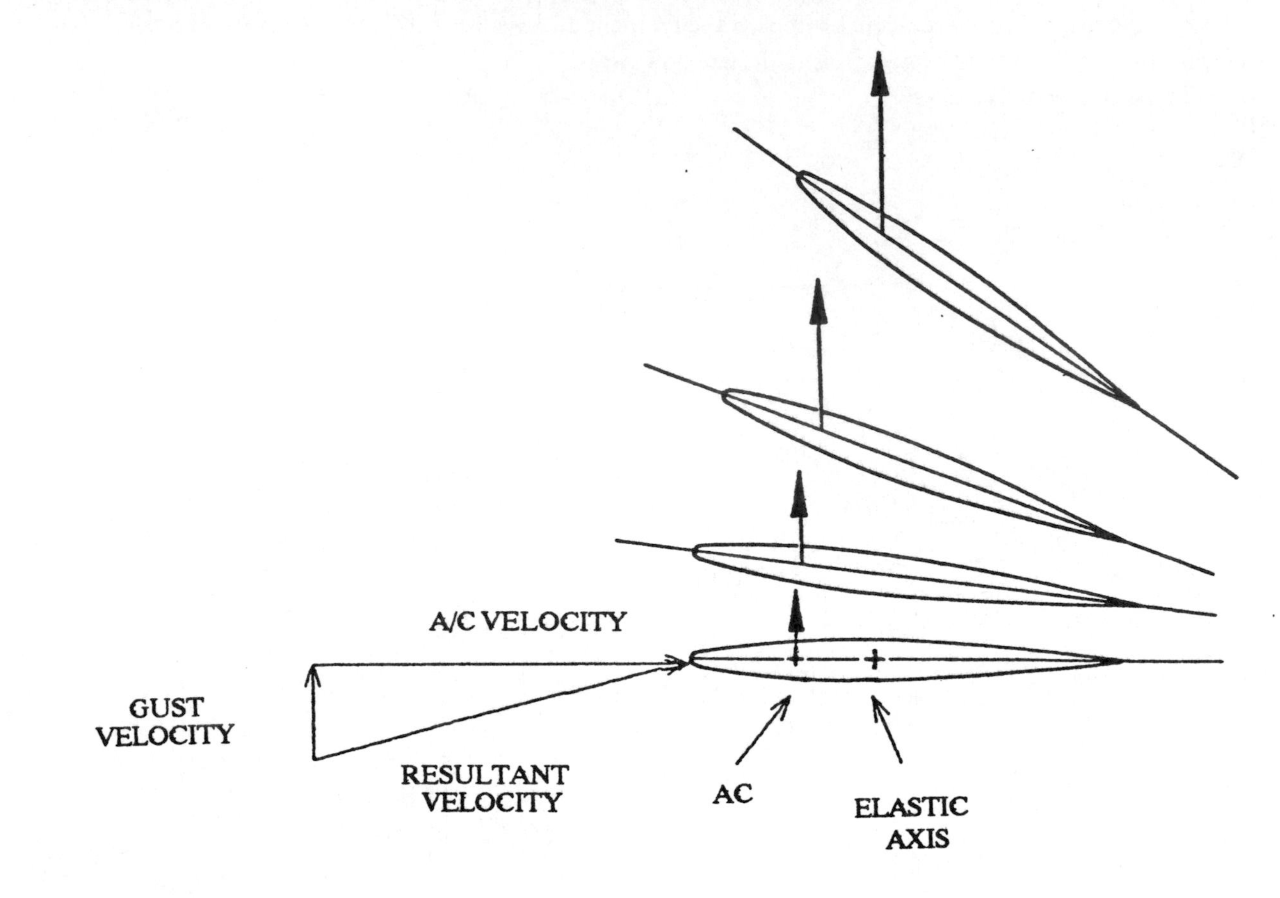


*Figure 1.4 Divergence*

*C. Flutter and Vibrations* are complex dynamic relationships involving aerodynamic forces, elastic deflections and inertia forces. The theory is far beyond the scope of this book. An extremely simplified explanation of the relationship between the various parts of the airplane will be attempted here. Each part of the airplane has its own natural frequency of vibration. This frequency can be mathematically calculated or measured by experiment. The interaction between two adjacent and connected parts can produce resonant vibrations if the natural frequencies are the same. A harmonic relationship to each other (half, third, twice etc.) produces a similar effect. This can be approximated by considering the deflection of a beam undergoing steady vibrational loads with no damping forces. The deflection, d, of the beam will be increased by a Dynamic magnification factor, Dmf. This relation is shown as:

$$d_{dynamic} = (d_{static})(Dmf)$$

The Dynamic magnification factor is:

$$Dmf = \frac{1}{1-(f_1/f_2)^2}$$

Where $f_1$ = natural frequency of the oscillating load
$f_2$ = natural frequency of the beam

Consider the case where $f_1$ is much less than $f_2$.
The $(f_1/f_2)^2$ term approaches zero and the magnification factor approaches 1.0. The beam deflection is the same as the static case.

Now consider the case where $f_1$ is much greater than $f_2$.
The $(f_1/f_2)^2$ term has a high value, but because it is in the denominator of the magnification factor, the factor itself approaches zero. Therefore the dynamic deflection approaches zero.

Finally, consider the case where $f_1$ and $f_2$ are equal.
The $(f_1/f_2)^2$ term now has a value of 1. The entire denominator is zero, the Dmf factor is infinity and the dynamic deflection approaches infinity. Resonance has been achieved and the beam oscillates with wild amplitude.

The vibration engineers have designed the airplane to avoid flutter and vibrations by keeping the natural frequencies of adjoining parts at different levels.

Maintenance personnel must be careful not to make unauthorized modifications of the aircraft structure. These may result in changes of the natural vibration frequencies.

## 3. Service Life.

MIL-A-008866B(USAF), Airplane Strength and Rigidity Reliability Requirements, Repeated Loads and Fatigue, states (in part):

"The design service life and design usage will be specified by the procuring activity in the contract. The design service life and design usage will be based on the mission requirements and will be stipulated in terms of:

a. Total flight hours.
b. Total number of flights.
c. total number and type of landings."

This Specification defines *durability* as: "The ability of the airframe to resist cracking (including stress corrosion and hydrogen induced cracking), corrosion, thermal degradation, delamination, wear, and the effects of foreign object damage for a specified period of time."

From the above definition of durability it appears that the subjects of Fatigue and Corrosion are of prime interest in prolonging service life. Much attention is devoted to these later, so they will not be discussed here.

## Creep

Although the specification does not specifically mention creep, the subject does definitely have a bearing on service life. Therefore it will be discussed here.

*Creep* can be defined as total strain (deformation) under constant load. The two factors, load and time, are required for creep to occur.

A third factor, heat, also may be required for appreciable amounts of creep. Most structural materials have sufficient creep resistance so that creep at normal temperatures is insignificantly small. At high temperatures creep in metals is important and must be avoided.

Turbine blades are extremely critical and the temperatures at which they operate is often the limiting factor in gas turbine operation. It is important that all cases of over temperature, such as hot starts, be reported to maintenance personnel so that they can check blade tip clearances.

The amount of creep depends upon the material, the amount of load being applied, temperature, and time. Figure 1.5 shows the effect of these factors.

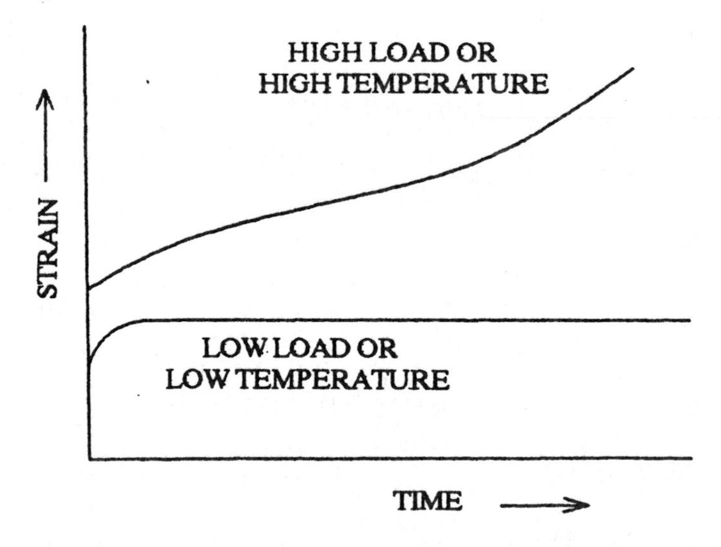


***Figure 1.5 Effects of load level, time, and temperature on creep.***

If the creep process continues until failure, three stages of creep take place as shown in Figure 1.6.

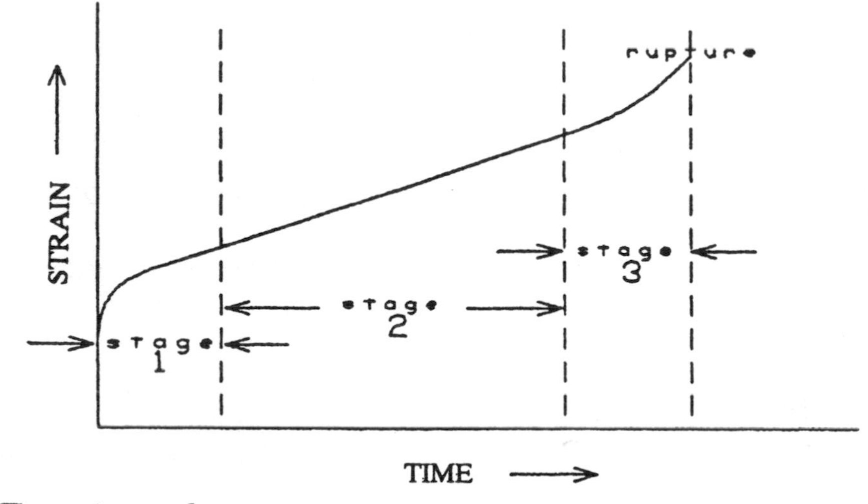


***Figure 1.6 Three stages of creep.***

When a tension load is first applied to a material, an initial strain, $\epsilon$, immediately occurs. This is called *primary creep.*

In stage 2, the creep rate continues at a constant rate. This is called steady-state or *secondary creep.*

In stage three, the rate shows a more rapid increase in rate until rupture occurs. This is called *tertiary* or *third stage creep.*

Fatigue, corrosion, and creep are all cumulative and irreversible. The damage will not "heal" with time. Corrective action must be taken as soon as damage is detected. Damage reports should be made promptly.

## EQUATIONS

1.1 $G - \frac{L}{W}$

1.2 $L - G \times W$

1.3 $LL - LLF \times W$

1.4 $ULF - 1.50 \times LLF$

## SYMBOLS

G = Load Factor, multiples of gravity
L = Load. Total lifting force, lb.
W = Weight of plane
LLF = Limit Load Factor, G units
LL = Limit Load, max. authorized, lb.
ULF = Ultimate Load Factor, Gs
UL = Ultimate Load, lb.

## REVIEW PROBLEMS

Note: Answers to problems are found at the end of the book (Page 165).

1. Static strength of an aircraft is its ability to withstand static loads.

   Static loads are: (a) gradually applied loads; (b) repeated loads; (c) rapidly applied loads.

2. "Limit load": (a) is the maximum weight of the airplane for takeoff; (b) is the load that, if exceeded, will cause failure of the airplane structure; (c) must be reduced if aircraft weight is increased; (d) is equal to the weight of the airplane multiplied by the limit load factor.

3. "Limit load factor": (a) is a fixed quantity for all flyable aircraft weights; (b) may be increased as weight decreases below design gross weight; (c) must be decreased as weight increases above design gross weight; (d) must be increased for asymmetrical flight.

4. The T-38a has a +LLF of 7.33 G's at a gross weight of 9600 pounds. If the gross weight is increased to 11,500 pounds, the LLF is: (a) 7.33; (b) 7.0; (c) 6.1; (d) 5.8.

5. "Rigidity" is: (a) assured if an airplane has adequate static strength; (b) important in the prevention of aeroelastic problems; (c) a factor in determining service life of an airplane; (d) a factor only in dynamic loadings.

6. Service life considerations include: (a) limit loads, ultimate loads, limit load factors; (b) aileron reversal, divergence, flutter; (c) fatigue, creep, corrosion.

7. Corrosion, fatigue and creep are all: (a) functions of time; (b) caused by repeated loads; (c) cumulative and irreversible.

8. Creep damage will most likely occur in a turbojet engine in the: (a) compressor section; (b) burner "cans"; (c) turbine blades; (d) afterburner section.

9. The two factors that are necessary for creep damage are: (a) heat and load; (b) time and heat; (c) load and time.

10. The three basic areas of aircraft design that we discussed are: (a) fatigue, creep and corrosion; (b) aileron reversal, divergence and flutter; (c) limit load factors, limit loads and ultimate load factors; (d) static strength, rigidity and service life.

# CHAPTER TWO

# LOADS ON AIRCRAFT

## 1. AIR PRESSURE LOADS

Air loads on an airplane are caused by differential static and dynamic air pressures. The relationship between static and dynamic air pressures is expressed by Bernoulli's equation. The sum of the static (non-moving) air pressure, P, and the dynamic (moving) air pressure, q, is considered
a constant. This is in agreement with the laws of the conservation of energy. "Energy can neither be created nor destroyed, but may be changed in form."

*Static Pressure* is the weight per unit of area of the air above the reference altitude.

*Dynamic Pressure* is similar to kinetic energy in mechanics and is calculated by the equation:

$$q = \frac{1}{2}\rho V^2 \quad (lb./ft.^2)$$

Where: $q$ = dynamic pressure, (lb./ft.$^2$)
$\rho$ = density of the air (slugs/cu.ft.)
$V$ = local velocity of the air(ft./sec.)

The air approaching the leading edge of an airfoil is first slowed. It then speeds up again as it passes over or beneath the airfoil.

Figure 2.1 shows a comparison of two local velocities with the flight path velocity, $V_1$, and with each other.

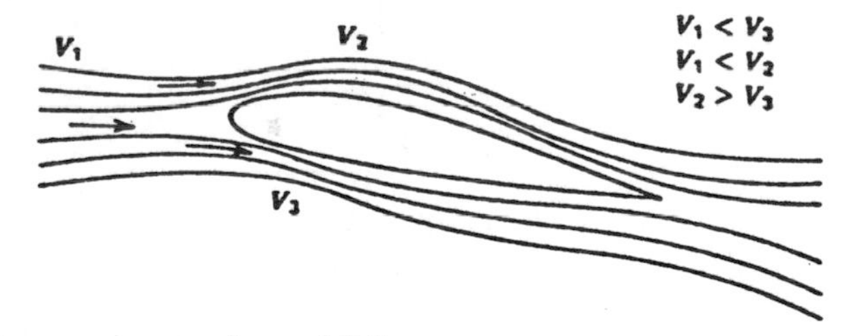


***Figure 2.1 Velocity changes around an airfoil.***

As the velocity changes, so does the dynamic pressure and, according to Bernoulli's equation, so does the static pressure. Air near the leading edge has slowed, and thus the static pressure in this region is higher than the ambient static pressure. Air that is passing above and below the airfoil, and thus speeds up to a value higher than the flight path velocity, will produce static pressure that is lower than ambient static pressure.

At a point near maximum thickness, maximum velocity and minimum static pressure will occur. Air pressure distribution about an airfoil that is producing lift is shown in Figure 2.2. Arrows pointing away from the airfoil show lower pressure than ambient and arrows pointing toward the airfoil show higher pressure than ambient.

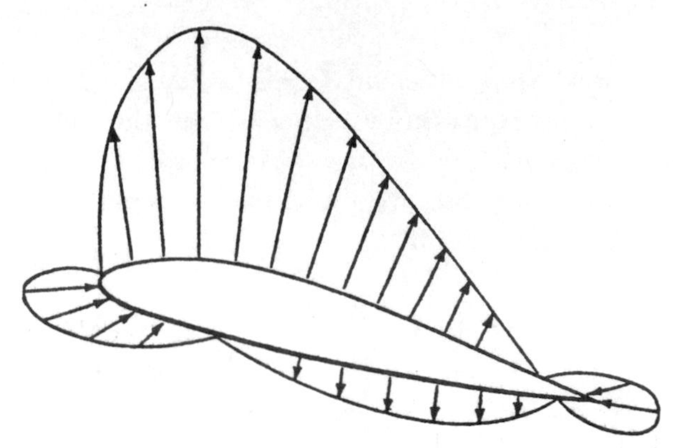

*Figure 2.2 Static pressure distribution.*

The resultant of the low and high pressure forces can be shown by a single vector. This is called the aerodynamic force, (AF), and is resolved into two component vectors. One component vector is parallel to the free stream velocity (relative wind) and is called *Drag.* The other is perpendicular to the relative wind and is called *Lift.*

These are shown in Figure 2.3.

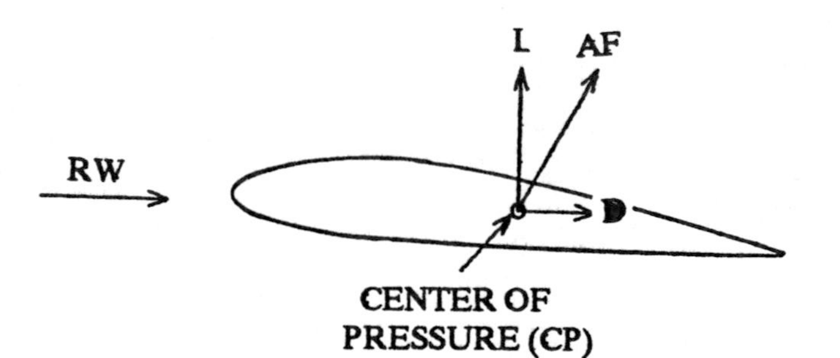


*Figure 2.3 Components of aerodynamic force.*

## 2. MANEUVERING LOADS

Maneuvering loads are those imposed by the pilot. One type, *general turning performance,* is common to all aircraft. It involves constant altitude turns and constant speed (but not constant velocity) conditions.

## General Turning Performance

In turning flight the airplane is **not** in a state of equilibrium, since there must be an unbalanced force to accelerate the plane into the turn. At this point let us review the subject of acceleration.

*Acceleration* is the change in velocity per unit of time. Velocity is a vector quantity that involves both the speed of an object and the direction of the object's motion.

Any change in the direction of a turning aircraft is therefore a change in its velocity, though its speed remains constant. An aircraft in a turn is acted upon by an unbalanced force toward the center of rotation.

Newton's second law states that an unbalanced force will accelerate a body in the direction of that force.

This unbalanced force is called the *centripetal force*, and it produces an acceleration toward the center of rotation known as *radial acceleration*.

The forces acting on an aircraft in a coordinated, constant altitude turn are shown in Figure 2.4.

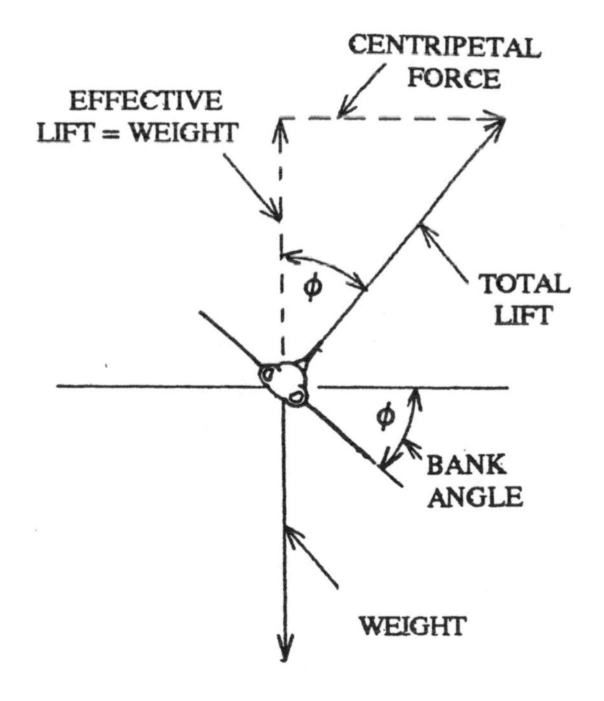


***Figure 2.4 Forces on aircraft in coordinated level turn.***

## *Loads on Aircraft*

In wings level, constant altitude flight, the lift equals the weight of the airplane and is opposite to it in direction.

In turning flight, the lift is not opposite, in direction, to the weight.

Only the vertical component of the lift, called effective lift, is available to offset the weight. Thus, if constant altitude is to be maintained, the total lift must be increased until the effective lift equals weight.

Solving the upper triangle in Figure 2.4:

$$\cos\phi = \frac{\textit{effective lift}}{\textit{total lift}} = \frac{W}{L} \tag{2.1}$$

The load factor G on the aircraft is lift/weight:

$$G = \frac{L}{W} \tag{2.2}$$

Inverting equation 2.1 and substituting the value of L/W into equation 2.2, we obtain:

$$G = \frac{1}{\cos\phi} \tag{2.3}$$

The *centripetal force* is found by solving the upper triangle in Figure 2.4:

$$\sin\phi = \frac{\textit{centripetal force}}{\textit{total force}}$$

So

$$\textit{centripetal force} = CF = L\sin\phi \tag{2.4}$$

The reaction force described in Newton's third law is called the *centrifugal force* and is shown in the bottom triangle in Figure 2.4.

A simplified vector diagram of the forces acting on an airplane in a coordinated constant altitude turn is shown in Figure 2.5.

From equation 2.2 it is seen that the total lift, L, can be replaced by its equivalent gross weight, GW.

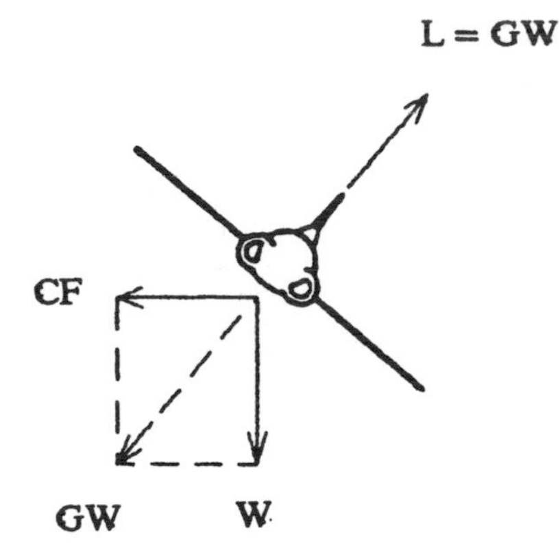


***Figure 2.5 Vector diagram of forces on aircraft in turn.***

## Load Factors on Aircraft in Coordinated Turn

Equation 2.3 tells us that the G's required for an aircraft to maintain altitude in a coordinated turn are determined by the bank angle alone. Type of aircraft, airspeed, or other factors have no influence on the load factor. Figure 2.6 depicts the load factors required at various bank angles.

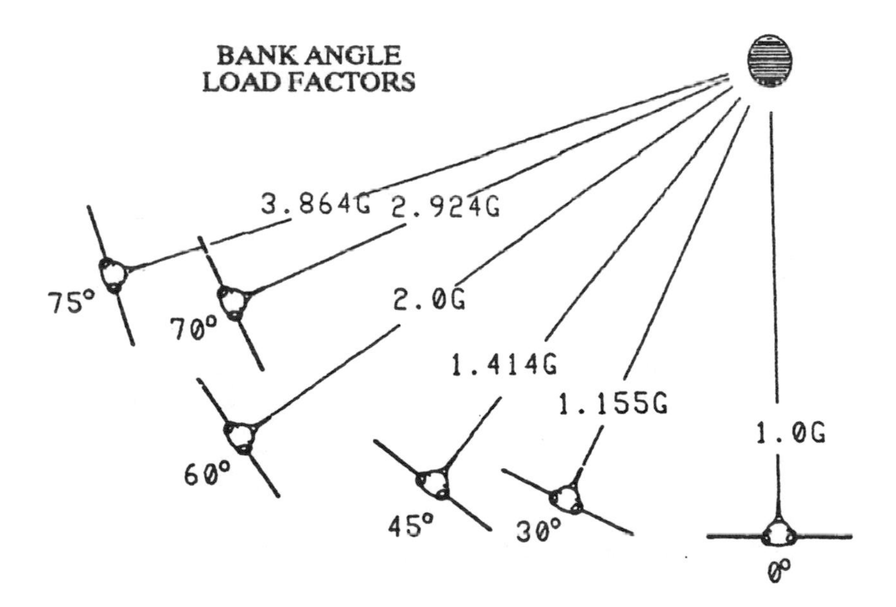

***Figure 2.6 Load factors at various bank angles.***

## *Loads on Aircraft*

Note that the load factors required to fly at constant altitude in turning flight were calculated by assuming that the wings provided all the lift of the aircraft.

According to this analysis, the load factor at a 90° bank will be infinity and thus impossible to attain. Yet we have all seen aircraft fly an eight point roll, with the roll stopped at the 90° position.

The secret is, of course, that lift is provided by parts of the aircraft other than the wings.

If the nose of the aircraft is held above the horizontal, the vertical component of the thrust will act as lift.

The fuselage is also operating at some effective AOA and also will develop lift. The vertical tail and rudder also will develop lift. Figure 2.7 shows the forces on an aircraft after it has rolled 90°.

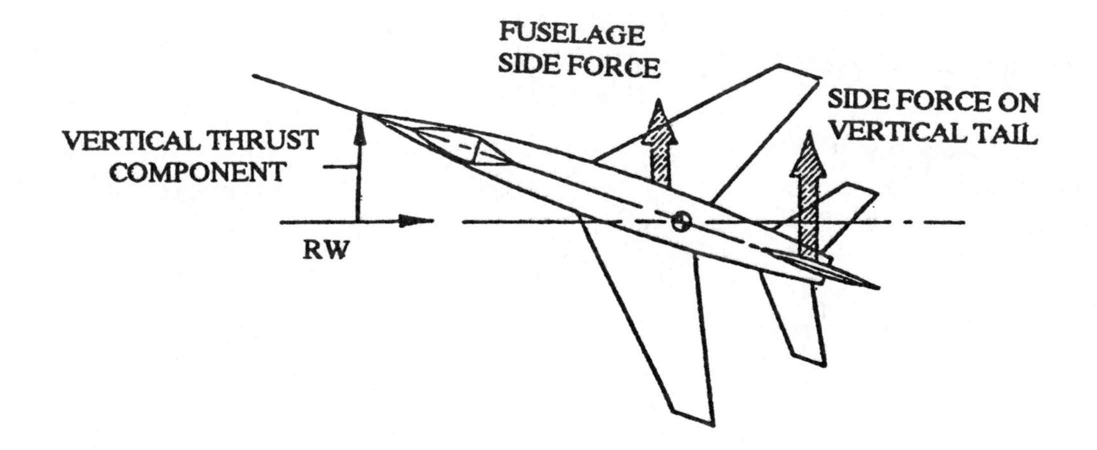


*Figure 2.7 Forces on aircraft at the 90° roll position.*

One of the three possible limiting factors on turning performance is the structural strength limits of the airplane. We have just discussed this limitation. The bank angle determines the load factor and may, therefore, limit the turn radius.

## Effect of Coordinated Banked Turn on Stall Speed

In basic aerodynamics we learned that the aircraft always develops its slow speed stall at the stall angle of attack (AOA). At this AOA the value of the lift coefficient is a maximum ($C_{L(MAX)}$). So stall speed ($V_S$) occurs at $C_{L(MAX)}$.

For 1 G wings level flight L = W and the basic lift equation can be written and solved for stall speed, $V_S$, as:

$$V_{S_1} = \sqrt{\frac{295W}{C_{L(max)}\sigma S}}$$

Since L = GW

$$V_{S_2} = \sqrt{\frac{295GW}{C_{L(max)}\sigma S}}$$

Dividing the second of the above equations by the first, shows that the stall speed depends on the square root of the G loading.

All other factors in the equation are constant for the same altitude and are thus canceled.

Therefore:

$$\frac{V_{S_2}}{V_{S_1}} = \sqrt{G} = \sqrt{\frac{W_2}{W_1}} \qquad (2.5)$$

Where $V_{S1}$ = stall speed under 1 G flight
$V_{S2}$ = stall speed under other than 1 G flight
G = load factor for condition 2

It should be noted that for the same value of $C_L$ (same AOA), the flight speed for all aircraft performance, **not only the stall speed**, also varies according to equation 2.5.

Substituting the value of G from equation 2.3 gives

$$\frac{V_{S_2}}{V_{S_1}} = \sqrt{\frac{1}{\cos\phi}} \qquad (2.6)$$

## 3. GUST LOADS

Air gusts impose both horizontal and vertical air loads on an aircraft in flight. Horizontal gusts impose side loads on the fuselage and vertical tail.

Maximum positive gust loads result when encountering vertical upward gusts.

These increase the effective angle of attack on the wing, thus increasing the lift on the wing and the wing loading.
This is shown in Figure 2.8.

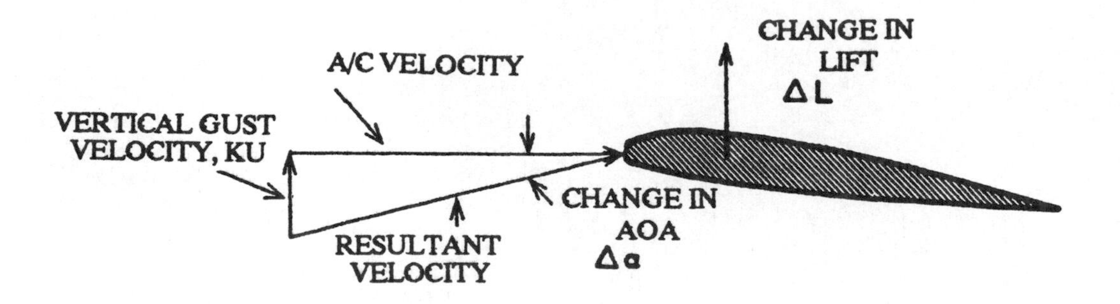


*Figure 2.8 Effect of vertical up gust.*

The increase in load factor due to an up gust is:

$$\Delta G = \frac{0.115 m \sqrt{\sigma}\, V_e (KU)}{W/S} \qquad (2.7)$$

Where G = increase in load factor due to up gust
m = slope of $C_L$-$\alpha$ curve
$\sigma$ = density ratio at flight altitude
$V_E$ = equivalent airspeed (knots)
KU = effective gust velocity (fps)
W/S = wing loading (psf)

The actual velocity of the up gust is U. In reality, the aircraft never feels this entire velocity instantaneously. Therefore U is multiplied by a gust alleviation factor, K, which is less than unity. In practice the maximum value of U, at sea level, for non-thunderstorm flying, is 50 feet per second. The value of K is about 0.6, although extremely sharp gusts could exceed this value.

It should be noted that high wing loadings (W/S) result in small increases in load factor. This leads to pilot complacency when flying in highly loaded aircraft. The pilot does not feel the turbulence and is not too concerned with it.

It must be remembered that the limit load factor is reduced as the weight goes up. The limit load factor goes down faster that the gust imposed load factor when the weight is increased. Therefore heavily loaded airplanes can be damaged without high G loadings. This is shown in Figure 2.9.

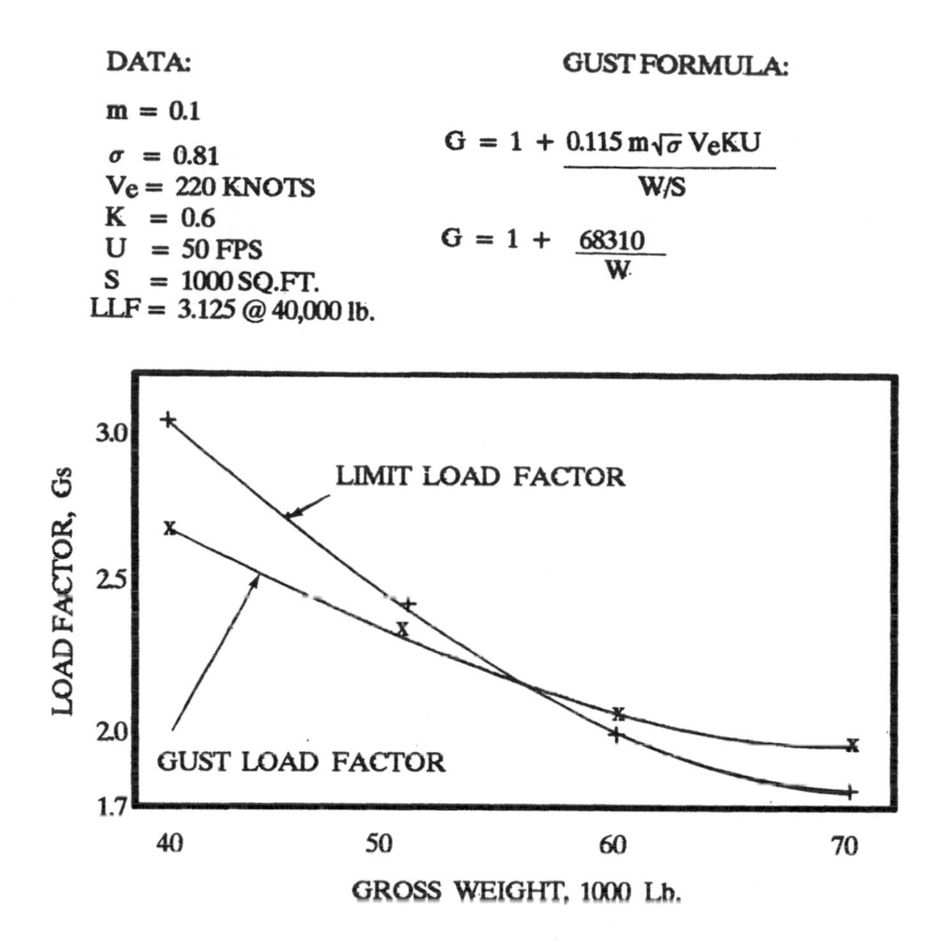


***Figure 2.9 Weight effect on LLF and on Gust Load Factor.***

## 4. LANDING AND GROUND LOADS

There are loads transmitted to the airframe during all ground operations. Heavy weight aircraft have potential dangers of tire or brake overheating. Horizontal loads imposed by wheel spin up, braking and rough landing surface are also significant, as are side loads caused by misaligned touchdowns and turning. The most critical loads occur during heavy weight, high sink-rate touchdowns. It is this case that we will explore further.

The landing gear must absorb the kinetic energy of the descending aircraft. On light weight aircraft this may be done with simple bungee springs or by simple spring steel gear struts. In heavier aircraft we generally rely on hydraulic shock absorbers, called *oleo struts*. These uniformly decelerate the rate of descent, (ROD), and the equations of constant acceleration can be applied.

## *Loads on Aircraft*

One constant acceleration equation is: $V^2 = 2as$.

Rewriting:

$$a = \frac{V^2}{2s} \tag{2.8}$$

Where: $a$ = vertical deceleration, $fps^2$.
$V$ = vertical velocity, (ROD), at touchdown, fps.
$s$ = effective strut deflection, ft.

The "G" forces on the aircraft are:

$$G = \frac{a}{g} \tag{2.9}$$

Where: $g$ = acceleration of gravity $\approx$ 32 $fps^2$.

Substituting (2.8) into (2.9):

$$G = \frac{V^2}{2gs} = \frac{(ROD)^2}{2gs} \tag{2.10}$$

There is generally no correlation between the actual load factor and what may be shown on the cockpit accelerometer as a response to landing loads. This is an inherent error in the instrument due to shock mounting and instrument response. Generally the instrument will overshoot and read far too high.

Limits on landing gear are determined by vertical rate of descent, not by load factors. USAF limits for fighter and attack type aircraft are 18.5 fps while USN limits are 22.5 fps. The higher navy limits are required for aircraft carrier operations.

Equation 2.10 shows the need to increase effective strut stroke to control landing loads if the design mission is to include high rate of descent tactics. This also should emphasize
the need for proper strut maintenance.

The equation also shows the need to control the rate of descent. Since this factor is squared, it follows that a 10 percent increase in ROD will increase the landing load factor by 21 percent.

Gross weight does not appear in the equation but it has two additive effects.

First, higher gross weight results in a higher force on the landing gear (F = GW). This is true even at the same load factor.

Second, increased gross weight dictates higher approach speed and, if flying the same glide path, a higher rate of descent results.

## ENGLISH SYMBOLS

a = acceleration, $ft/sec^2$
AF = aerodynamic force, lb.
$C_{L(MAX)}$ = maximum lift coefficient
CP = center of pressure location
D = drag, lb.
F = unbalanced force, lb.
g = acceleration of gravity
G = n = load factor, dimensionless
K = gust alleviation factor
L = lift, lb.
m = mass, slugs
m = slope of $C_L$-$\alpha$ curve
q = dynamic air pressure, psf
ROD = rate of descent, fps
RW = relative wind direction
s = distance, ft.
S = wing area, sq.ft.
U = gust velocity, fps
W = weight, lb.

## GREEK SYMBOLS

$\alpha$ (alpha) = angle of attack, degrees
$\Delta$ (DELTA) = "change in"
$\rho$ (rho) = density of air, $slugs/ft^3$
$\sigma$ (sigma) = density ratio
$\phi$ (phi) = bank angle, degrees

## EQUATIONS

2.1 $\cos\phi = \frac{W}{L}$

2.2 $G = \frac{L}{W}$

2.3 $G = \frac{1}{\cos\phi}$

2.4 $CF = L\sin\phi$

2.5 $\frac{V_{S_2}}{V_{S_1}} = \sqrt{G} = \sqrt{\frac{W_2}{W_1}}$

2.6 $\frac{V_{S_2}}{V_{S_1}} = \sqrt{\frac{1}{\cos\phi}}$

2.7 $\Delta G = \frac{0.115 m \sqrt{\sigma}\, V_e (KU)}{W/S}$

2.8 $a = \frac{V^2}{2s}$

2.9 $G = \frac{a}{g}$

2.10 $G = \frac{V^2}{2gs} = \frac{(ROD)^2}{2gs}$

## Review Problems

1. A 10,000 lb. airplane is making a 50° banked, level altitude turn.
Calculate the Gs on the airplane and the centripetal force.
For 50°, sin = 0.766, cos = 0.643, tan = 1.192.
1.56 G , 11,913 pounds.

2. The above airplane has a wings level stall speed of 120 knots at 10,000 lb. gross weight.
Find its stall speed in a 60° level bank.
For 60°, sin = 0.866, cos = 0.5, tan = 1.732.
170 Knots

3. Two identical cargo airplanes are flying in formation. Plane A has a full load of cargo. Plane B is empty.
Which plane will have the higher "G" reading upon encountering turbulent air?
plane B

4. Which of the above airplanes will is more likely to be damaged by the wind gusts?
plane A

5. An airplane lands after making an instrument landing without flaring the airplane. The ROD was 900 feet per minute at touchdown. The shock strut stroke for the plane is six inches. Assuming constant deceleration, calculate the actual "G" loads on the plane.
7 G's

# CHAPTER THREE

# V-G (V-n) DIAGRAM (FLIGHT ENVELOPE)

Equation 2.5 shows that the stall speed varies as the square root of the G loading applied to an aircraft. If an aircraft is under a 4 G load, the stall speed is the square root of 4 or twice the 1 G stall speed. At zero G the stall speed is zero because no lift is developed. This formula cannot be applied when an aircraft is developing negative load factors (square root of a negative number is impossible). But it is possible to stall the aircraft under negative G loading.

Figure 3.1 shows the first construction lines of the V-G diagram (Flight Envelope), the plot of the stall speed at various G loadings. These curved lines also can be thought of as the number of G's that can be applied to the aircraft before it will stall at any airspeed.

They are called the *aerodynamic limits* of the aircraft. It is impossible to fly to the left of these curves, because the aircraft is stalled in this region.

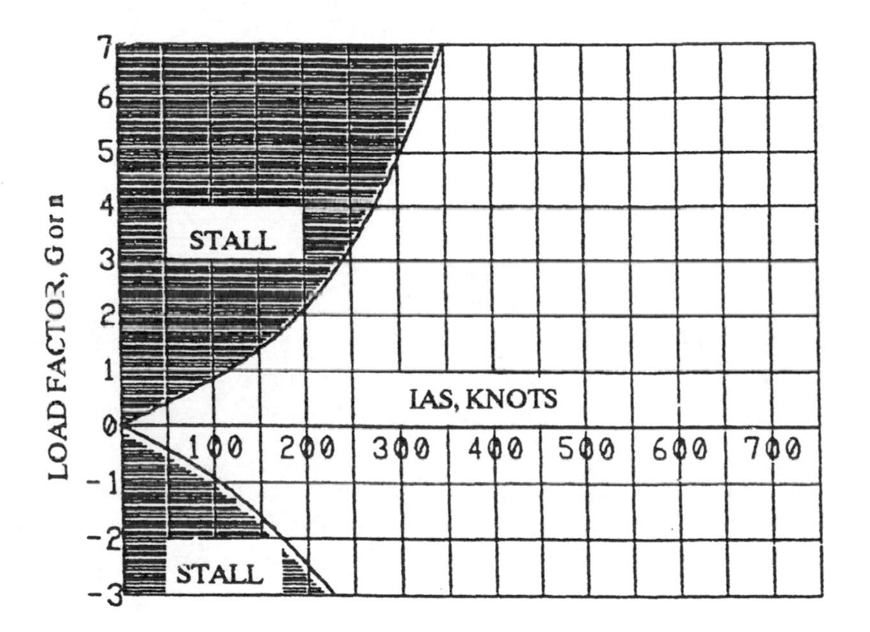


***Figure 3.1 First stage construction of V-G diagram.***

The second stage of construction of the V-G diagram has horizontal lines at the positive and negative limit load factors, LLF. These limits are specified for each model aircraft and are shown in Figure 3.2 as "MAXIMUM DESIGN 7G" and as "MAXIMUM DESIGN-3G."

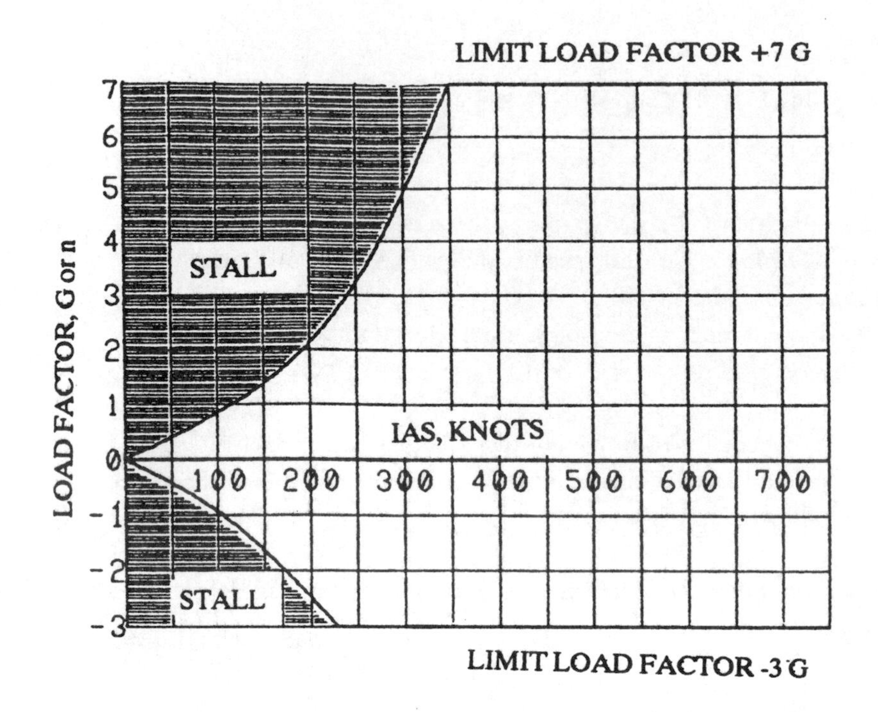


***Figure 3.2 Second stage construction of V-G diagram.***

## Maneuver Speed

An interesting point on the V-G diagram is the intersection of the aerodynamic limit line and the structural limit line. The aircraft's speed at this point is called the *maneuver speed,* $V_p$, commonly called the corner speed. At any speed below this speed the aircraft cannot be overstressed, as it will stall before the limit load factor is reached. Above this speed the aircraft can exceed the limit load factor before it stalls. At the maneuver airspeed the aircraft's limit load factor will be reached at the lowest possible airspeed.

Because the bank angle determines the load factor in a level turn, the maximum permissible bank angle and the lowest possible airspeed are at $V_p$. Thus minimum radius of turn will result. Maneuver airspeed is also the best speed to penetrate turbulence, as it is impossible to exceed limit load factor at this speed. The maneuver airspeed is shown in Figure 3.3.

Maneuver speed can be calculated by

$$V_P - V_S \sqrt{LLF} \qquad (3.1)$$

$V_p$ = maneuver speed (knots)
$V_s$ = stall speed (knots)
LLF = limit load factor

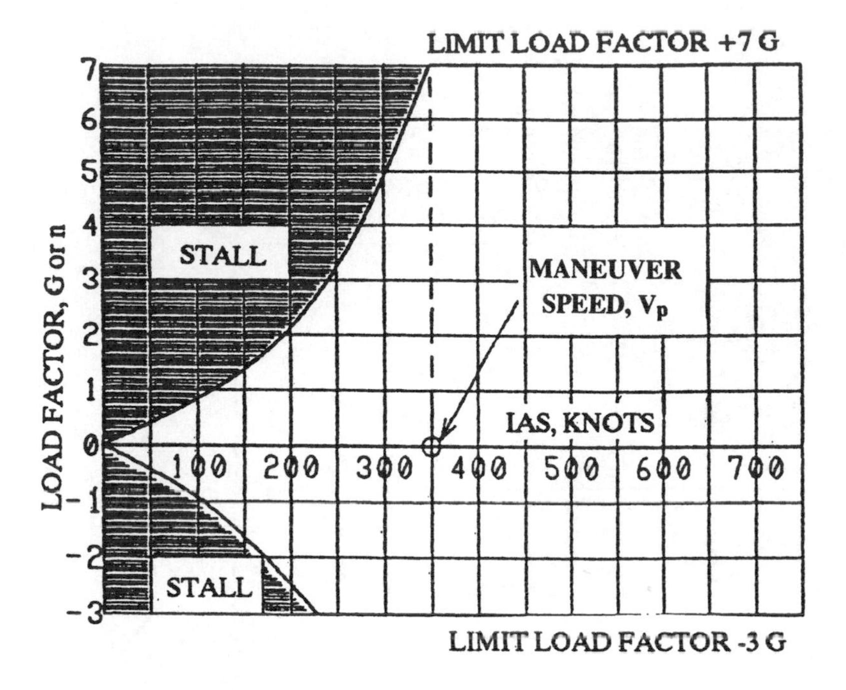


***Figure 3.3 Maneuver (corner) speed.***

All proposed missions of an aircraft can be performed without exceeding the limit load factors, but on occasion a pilot will overstress the aircraft. If this happens, the aircraft probably will suffer permanent objectionable deformation, resulting in costly repairs, but it will not necessarily result in failure of the primary structure.

There is another load factor that, if exceeded, may lead to catastrophic failure of the airframe. This load factor is called the ultimate load factor, ULF. Numerically the positive and negative ultimate load factors are 1.5 times the limit load factors. These are shown in Figure 3.4 and are labeled as "STRUCTURAL FAILURE."

## HIGH SPEED LIMIT

The high speed limit of the V-G diagram is called the *Limit Speed,* ($V_L$) or, more commonly, the redline speed. It is shown in Figure 3.4. If this speed is exceeded one of the following problems may be encountered

1. Aeroelastic problems (aileron reversal, divergence, flutter).
2. Structural failure due to high dynamic pressure (high q).
3. Compressibility problems for non-supersonic aircraft.
4. Loss of directional stability at high Mach numbers.
5. Aerodynamic heating problems (see Figures 3.5 and 3.6)

### *V-G Diagram*

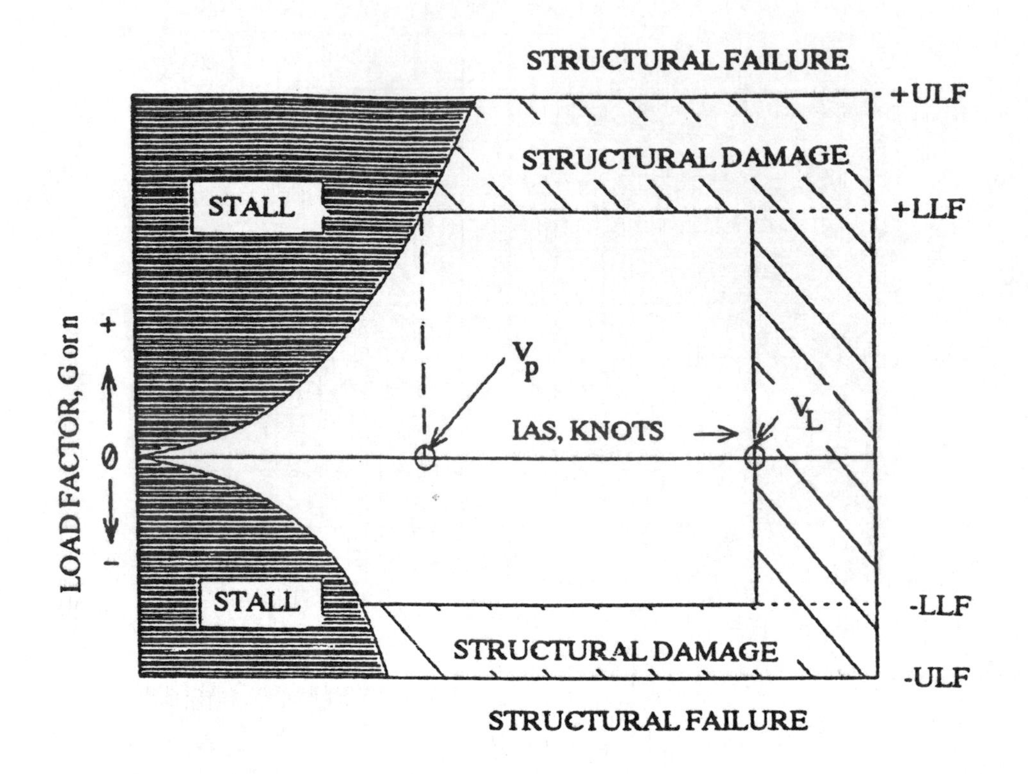


*Figure 3.4 Ultimate load factors.*

Aerodynamic heating problems result from two different causes.

The first problem is the heating of the air at the engine intakes. All jet engine compressors are designed to operate subsonically. When the airplane is flying supersonically the air must be slowed to subsonic speed. This occurs when the air passes through one or more shock waves. Each shock wave slows the air, compresses and heats it. The faster the airplane flies, the more the air is heated and, consequently, the engine produces less thrust.

The second type of aerodynamic heating is produced by the air being compressed at the leading edges of wings and tail surfaces. These regions are known as stagnation points. This type of heating weakens the metal surfaces. Figures 3.5 and 3.6 illustrate the effects of aerodynamic heating on a Mach 2 and a Mach 3 supersonic airplane. Figure 3.5 shows that the stagnation temperature of the Mach 2 aircraft is about 250°F while the Mach 3 aircraft has a stagnation temperature of about 650°F.

Compare the heating effect on the Ultimate Tensile Strength of various metals as shown in Figure 3.6. All three of the metals have sufficient strength to be used for the leading edges of the Mach 2 airplane. But, both aluminum alloy and titanium alloy lack the strength and are not suitable for the Mach 3 aircraft.

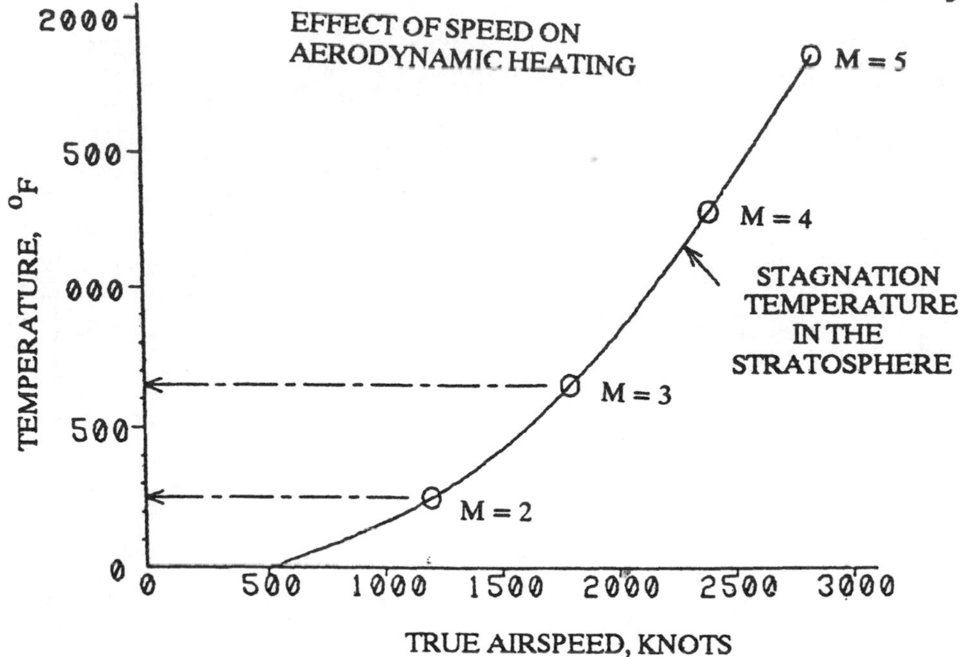


*Figure 3.5 Stagnation temperatures.*

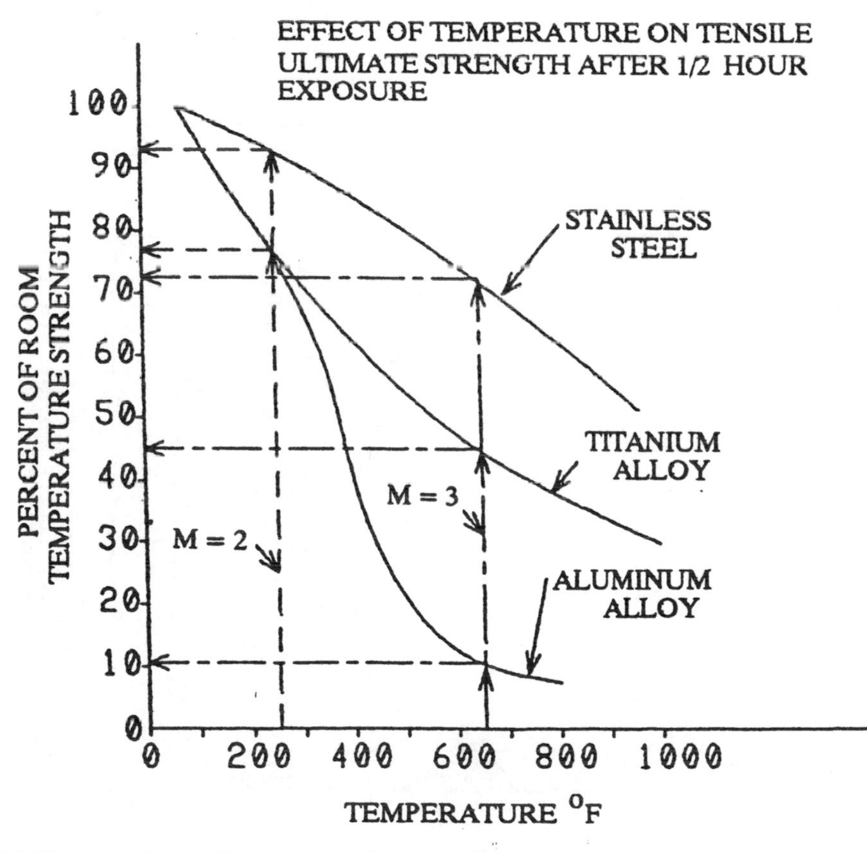


*Figure 3.6 Temperature effect on tensile strength.*

*V-G Diagram*

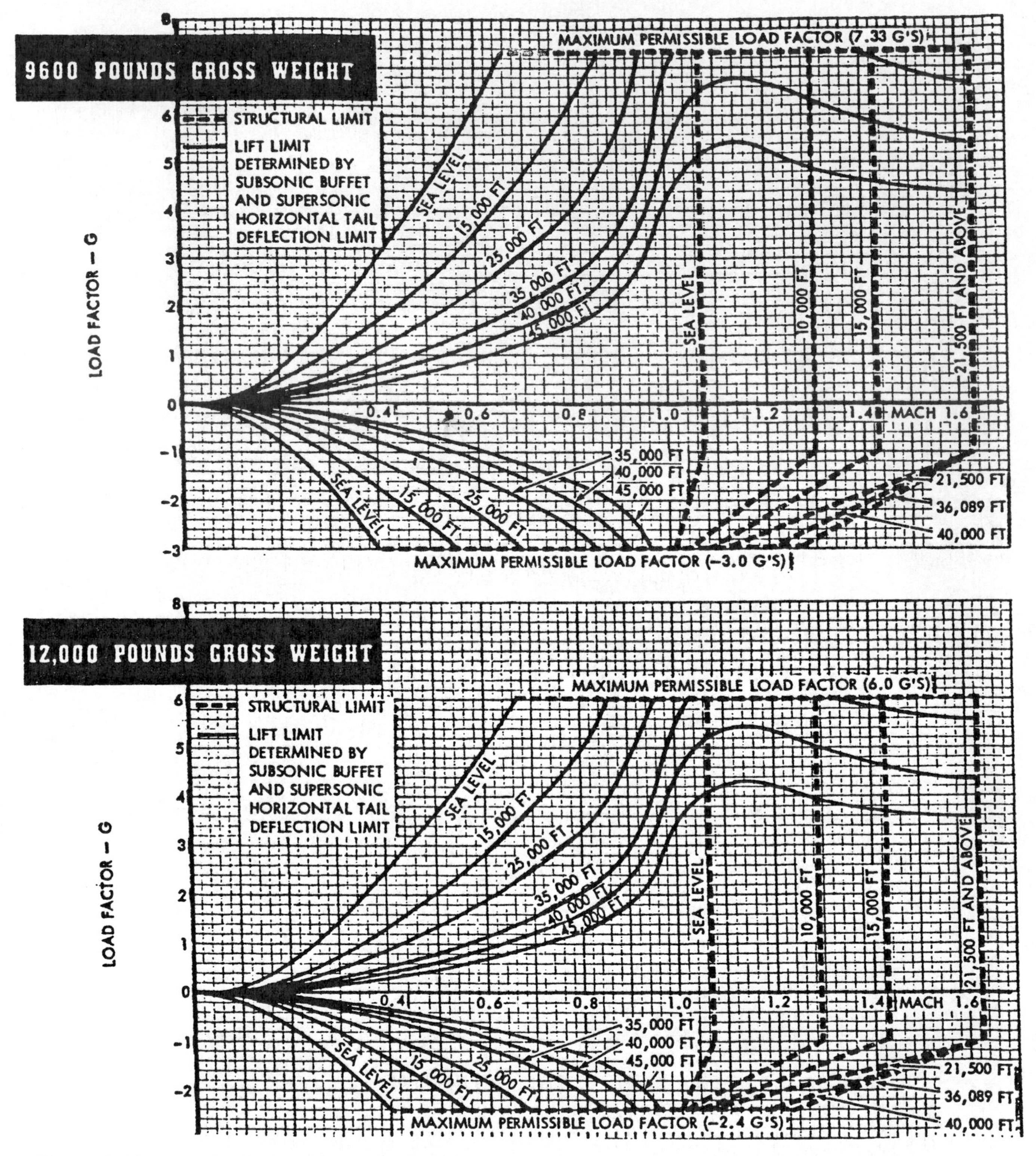


*Figure 3.7 Symmetrical V-G diagrams for T-38A.*

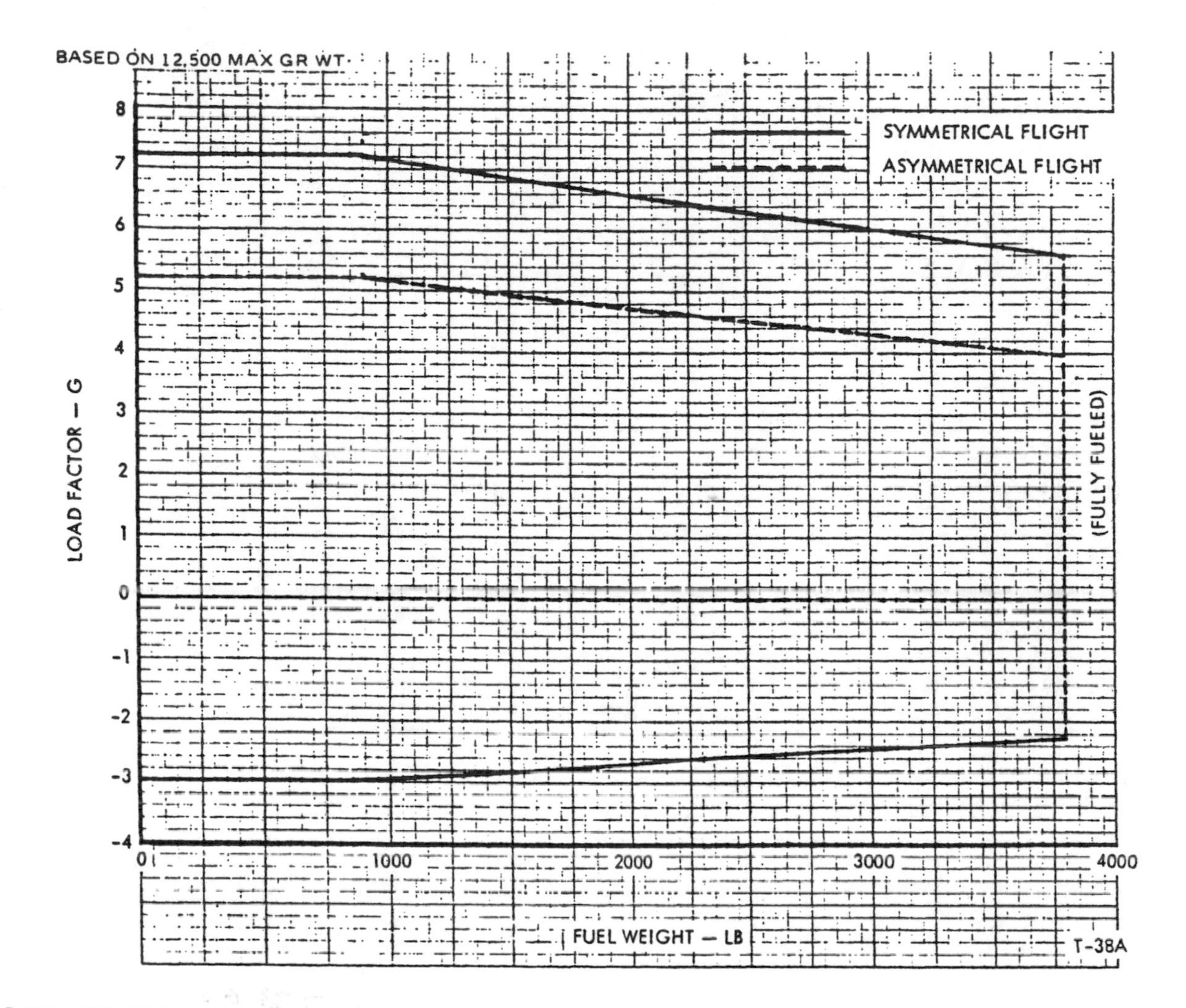


## LOAD FACTOR LIMITATIONS.

Do not exceed the following

**SYMMETRICAL FLIGHT.**

| *Load Factor (G's)* | *Weight of Fuel Remaining (Pounds)* |
|---|---|
| -2.3 to +5.6 | 3800 |
| -2.6 to +6.4 | 2300 |
| -3.0 to +7.33 | 900 or less |

**ASYMMETRICAL FLIGHT.**

| *Load Factor (G's)* | *Weight of Fuel Remaining (Pounds)* |
|---|---|
| 0 to +4.0 | 3800 |
| 0 to +4.6 | 2300 |
| 0 to +5.22 | 900 or less |

*Figure 3.8 Load factor limitations for T-38A.*

*V-G Diagram*

## SYMBOLS, FORMULAS, PROBLEMS

## ENGLISH SYMBOLS

LLF = limit load factor, "G" units
ULF = ultimate load factor
$V_L$ = limit airspeed, knots or Mach number
$V_p$ = maneuver airspeed
$V_s$ = 1 "G" stall airspeed

## EQUATION

3.1 $V_P - V_S \sqrt{LLF}$

## REVIEW PROBLEMS

From Figure 3.7, for a 9600 lb. T-38A:

1. Find the maximum number of +"Gs" that can be pulled at $V_p$ at 15,000 ft. 7.33 G's

2. Find the $V_L$ at 10,000 ft. 1.3 m.

3. Can the airplane exceed +LLF at any airspeed at 40,000 ft.? NO

4. How many +"Gs" can be pulled at sea level and 0.6 Mach? 6.2 G's

5. What is the +3 "G" stall speed at 15,000 ft.? .55 m

# CHAPTER FOUR

# METALLIC COMPOSITION

## THE ATOM

The structure of materials on the micro level consists of atoms, molecules and crystal structures.

Atoms are so small that they cannot be seen even with the most powerful electron microscope. Atoms are made up of even smaller particles. Think of an atom as a miniature solar system. The central core of the atom is called the nucleus. This can be compared to the sun in our solar system. The nucleus has several kinds of particles that are packed closely together.

The two basic particles in the nucleus are the *protons* and the *neutrons*. The protons are positively charged particles. All atoms of the same element have the same number of protons, but this number is different for each element. The neutrons have no electrical charge. The proton and the neutron have the same size and mass. Nearly all the weight of the atom is concentrated in the nucleus.

The remainder of the atom is made up of an electron field. *Electrons* are negatively charged particles. They whirl around the nucleus at great speed. In our comparison of the atom to the solar system, it is convenient to think of the electrons as planets in orbit about the sun. There are the same number of electrons as protons, so the entire atom is electrically neutral. The mass of the proton is about 1,836 times as great as the electron.

Figure 4.1 shows a schematic of an atom.

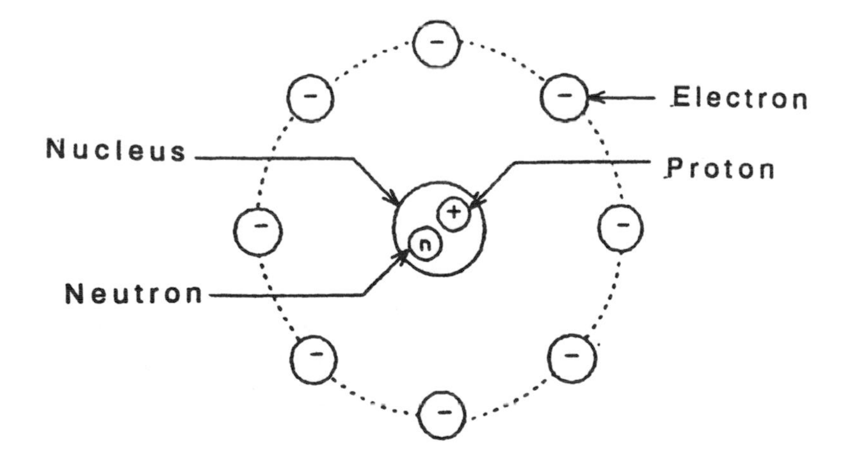


***Figure 4.1 Main parts of the atom.***

### *Metallic Composition*

The atom is mostly space. Picture a child's playing marble as the nucleus of an atom. If this marble was at the center of a football field, the outer electrons would be at the goal posts.

## METALLIC ONDING

There are several methods (depending upon the type of atoms) by which atoms can bond together to form solid materials. As we are interested in metals used in aircraft construction, the type of bonding that is pertinent is called *metallic bonding*.

In metallic bonding the atoms have only a few electrons in their outer shells. These are called *valence electrons*. They are "free" electrons that detach from the atom and form an *electron cloud*. This changes the atom to an *ion*, which is positively charged. Thus the attracting force between the unattached electrons and the ions bond the atoms together. It is the resistance to these bonds to external forces that gives metals their strength.

Figure 4.2 shows the metallic bonding process.

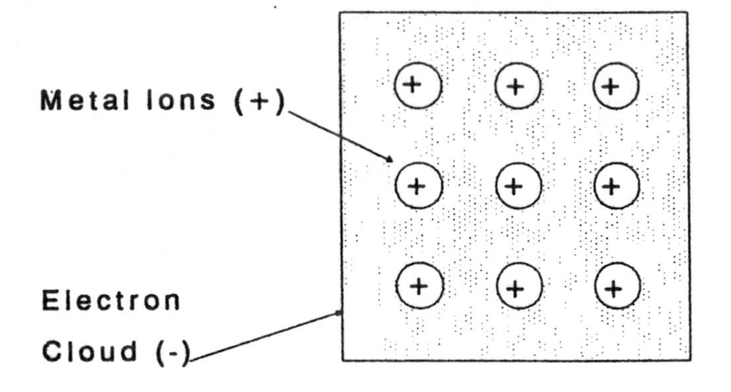


***Figure 4.2 Metallic bonding.***

## KINETIC THEORY OF MOLECULAR MOTION

The kinetic theory explains the effect of heat on matter. Two basic assumptions must be made. First that matter is composed of submicroscopic particles (atoms, ions, or molecules). Second, these particles are in constant motion.

The speed of the particles is dependent on the temperature. As the temperature increases, the particles speed increases also. If the temperature of a solid metal is raised to a point where the velocity of the atoms is so great that the atomic bonds break and the metal becomes a liquid. Conversely, upon lowering the temperature of a liquid metal, the velocity of the atoms decreases, the atomic bonding reforms and the metal "freezes."

## CRYSTAL UNIT CELLS

Structural metals in the solid state form from the liquid state as *crystals.* A crystal is a rigid body in which the constituent particles are arranged in a repeating pattern. The basic building block of the crystal is known as a unit cell. The crystal is built from the repetition of these identical unit cells.

This is similar to the building of a brick wall (the crystal) using identical bricks (the unit cells). The shape and number of atoms in the unit cell depend upon the metal. There are seven different crystal unit cells, but structural metals used in aircraft consist of only three of these. They are called the *body centered cubic,* the *face centered cubic,* and the *hexagonal close packed.*
These are shown schematically in Figure 4.3

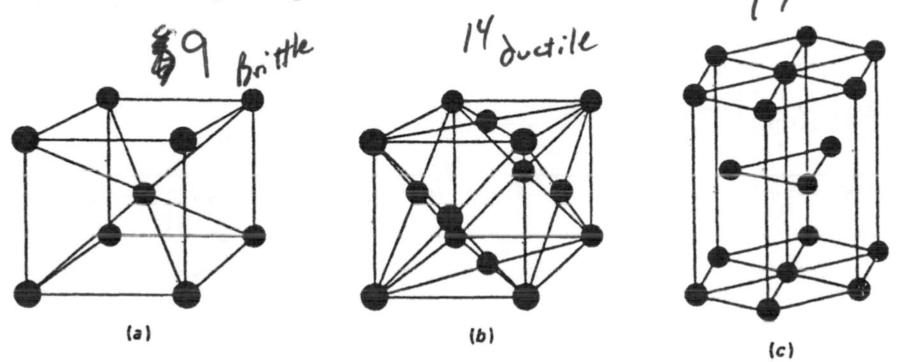


***Figure 4.3 Common unit cells: (a) body centered cubic; (b) face centered cubic; (c) hexagonal close packed.***

The body centered cubic (BCC) has a total of nine atoms. One is at each corner of the cube and one in the center.

Metals which combine in the body centered cubic cell include Chromium, Molybdenum, Tungsten and Iron (below 910°C). Generally, these are brittle materials.

The face centered cubic (FCC) unit cells consist of 14 atoms. One atom is at each cube corner and one is in the center of each face. Aluminum, Copper, Gold, Nickel, Silver and Iron (above 910°C) are examples of metals that have the FCC form. These are ductile metals.

Cobalt, Magnesium, Titanium and Zinc have the hexagonal close packed (HCP) arrangement. There are 17 atoms in an HCP unit cell.

The complete crystal is a lattice made up of unit cells built up in three dimensions as shown schematically in Figure 4.4.

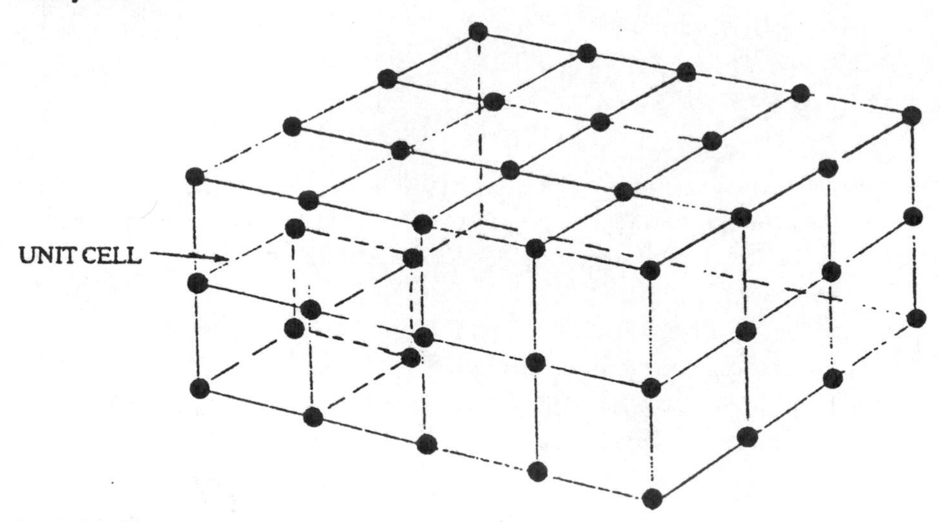

*Figure 4.4 Crystal lattice structure.*

## METALLIC SOLIDIFICATION

The process of reestablishing atomic bonds upon cooling of a liquid metal is not instantaneous, but starts at certain nuclei and grows in all directions in a dendritic (branch-like) manner. When the growth process is interrupted by interference with other dendrites from adjoining nuclei, a grain boundary is formed.

The solidification of pure copper is shown in Figure 4.5.

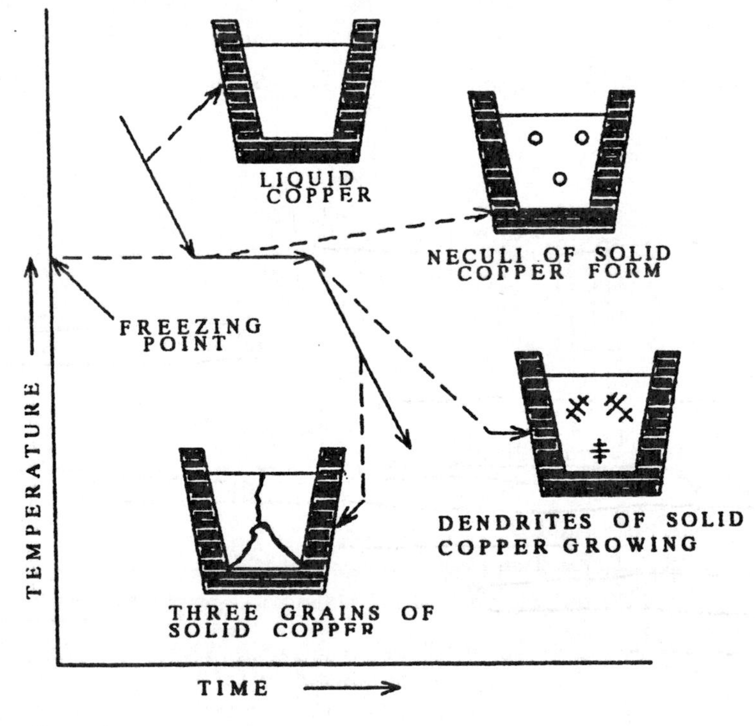


*Figure 4.5 Solidification of copper.*

In the freezing process the nuclei are oriented in random directions. The crystalline structure of adjacent grains are not necessarily aimed in the same direction. This is shown in Figure 4.6. The grain boundaries are very important to the characteristics of the metal. This will be discussed soon.

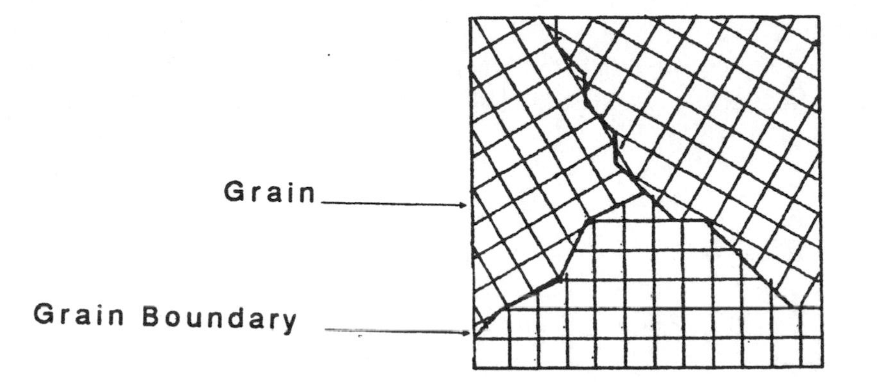


*Figure 4.6 Grains and grain boundaries.*

Grain size is influenced by several factors. Among these are: the type of metal; rate of cooling; subsequent heat treatment; and the direction of the grain of the material. Generally, the more ductile metals have larger grains. Slow rate of cooling also produces large grains.

In casting, the metal close to the mold surface cools faster than the interior. This not only produces smaller grains near the mold surface, but also produces columnar grains rather than equiaxed grains in this region. This is shown schematically in Figure 4.7.

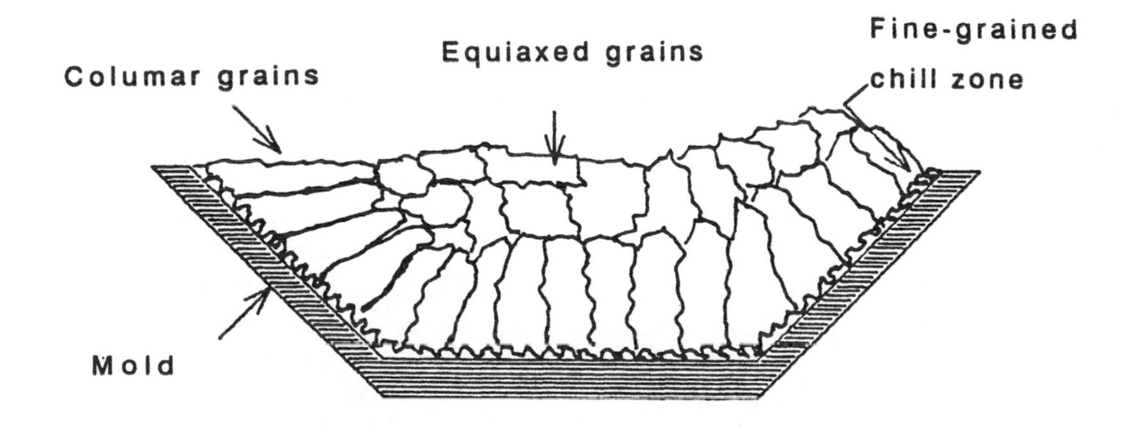

*Figure 4.7 Typical ingot structure.*

Between the grains there are regions known as grain boundaries. Grain boundaries can either increase or decrease the strength of a metal. Some grain boundaries contain a precipitation of second-phase particles that chemically bind the grains together. This is similar to mortar holding bricks together. This strengthens the metal. On the negative side, grain boundaries often contain impurities that can lead to stress corrosion. The boundaries may be composed of brittle materials such as iron carbides that prevent the metal from retaining the flexibility required in aircraft.

## CRYSTAL IMPERFECTIONS

Commercial metals are not nearly as strong as the atomic bonding theory predicts as the crystals are far from perfect. There are many imperfections within the crystal lattices. These can be divided into two principal types: *point defects* and *line defects*, known as *dislocations.*

### Point Defects

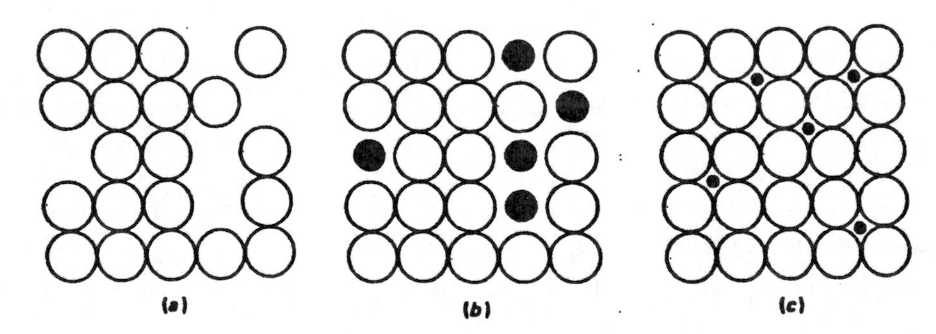


***Figure 4.8 Point defects in crystal lattices: (a)vacancies, (b)substitutions, (c) interstitial atoms.***

*Vacancies* are missing atoms in the lattice structure. If vacancies are present the remaining atoms can move around into the holes. The vacancies themselves can join up with other vacancies and thus a form a crack.

*Substitution* occurs when foreign atoms fill the vacancies. If impurity atoms are substituted, the metal will be weakened. Alloying of metals involves the substitution process and here improvement of the base metal properties results.

*Interstitial atoms* may be the same atoms as in the crystal or they can be foreign atoms. In either case, the extra atoms distort the lattice, create internal stresses, increase the strength and hardness, but decrease the ductility of a metal.

## Dislocations

Crystal *dislocations* are the major reason that the actual strength of metals is so much less than the theoretical strength. A dislocation is a line of defects and is an interruption of the regular arrangement of atoms in the lattice structure. There are several types of dislocations but, for simplicity, only the edge dislocation shown in Figure 4.9 will be discussed.

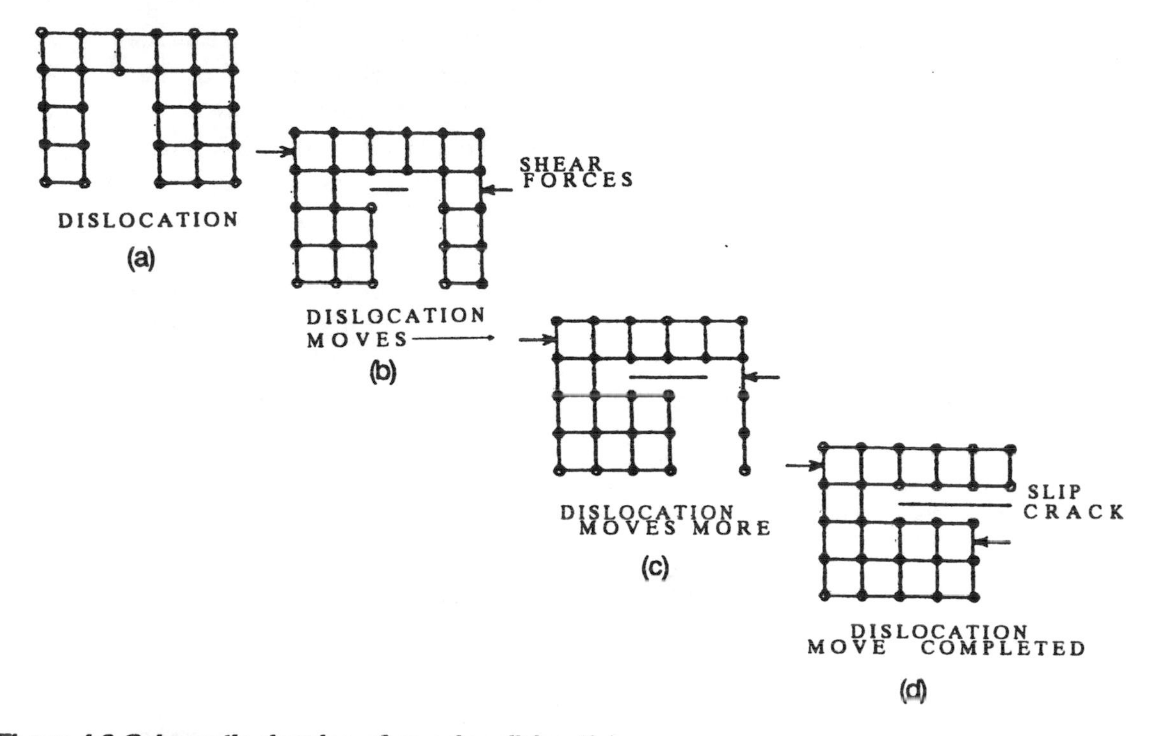


***Figure 4.9 Schematic drawing of an edge dislocation.***

Most authors describe a dislocation as a row of extra atoms inserted in the crystalline lattice. But, it seems that a row of missing atoms would describe the same condition.

Figure 4.9(a) shows a two dimensional schematic of a dislocation. When sufficient shear forces are applied, as shown in Figure 4.9(b) the atoms at the lower right of the dislocation have less resistance to moving to the left because of the absence of intervening atoms.

The row of atoms to the right of the dislocation then moves to the left, thus closing the dislocation gap.

This, in effect, moves the dislocation to the right replacing the row of moved atoms. The bond is broken between the atom that moves and the atom above it, that does not move.

This produces a sub-microscopic crack known as a *slip line*. The moving process is then repeated with the next row of atoms and the dislocation continues to move and the slip line elongates.

Finally the dislocation stops moving because the shearing load is removed or because the dislocation hits a barrier to its movement. The barrier is usually a grain boundary. The dislocation is now removed from the crystalline lattice (see Figure 4.9(d)) and further atomic movement is resisted by the perfect arrangement of atoms.

While the dislocations are moving a process known as slip is occurring. The stress in the material is reduced during slip as the metal is deforming more rapidly than the shearing force is being applied. The amount of slip depends upon the number of dislocations in the metal. Generally, ductile materials have more dislocations than brittle metals. The grain size of ductile metals is also larger, so the dislocations can move farther before being stopped by a grain boundary. It has been estimated that a one centimeter cube of a ductile material has $10^8$ dislocations.

Although the strength of a metal increases when the dislocations have closed, a lasting damage remains from the slip lines. These combine with other slip lines and when a thousand or so join a larger crack, called a slip plane, forms. Slip planes can sometimes be seen with the naked eye. They are the potential plane of shear failure, but also they may be the start of a fatigue failure. This will be discussed in more detail later.

## REVIEW PROBLEMS

1. Magnesium and Titanium unit cells form as: (a) BCC; (b) FCC; (c) HCP.

2. Aluminum, Nickel and Copper unit cells form as: (a) BCC; (b) FCC; (c) HCP.

3. Steel (at room temperature), Chromium, and Tungsten form as: (a) BCC; (b) FCC; (c) HCP.

4. If an entire row of atoms are missing, from an atomic lattice, this is called a: (a) vacancy; (b) interstitial; (c) dislocation; (d) substitution.

5. A foreign atom located between the spaces of the parent atoms in an atomic lattice is called a: (a) interstitial; (b) dislocation; (c) substitution; (d) vacancy.

# CHAPTER FIVE

# ELEMENTARY STRESS ANALYSIS

## PROPERTIES OF AIRCRAFT MATERIALS

Load forces on aircraft structures cause the members to stretch, compress, bend, buckle, break, split, or crush. Before we study the effects of the loads on the structure, we must understand some definitions of the properties of the materials.

### Strength

The *strength* of a material is its ability to withstand externally applied forces without failing. When a part can no longer perform its function, it has failed. This can be either a breakage of the part or a deformation to the point where it can not do its job.

### Elasticity

*Elasticity* is the ability of a material to return to its original size and shape after applied loads have been removed. If the loads on the material become excessive it will not return to its original size and shape. In this case the *elastic limit* of the material has been exceeded and a permanent deformation, called permanent *strain*, results.

### Plasticity

*Plasticity* is the opposite of elasticity. Elastic materials, when they are deformed beyond their elastic limit undergo plastic deformation.

### Ductility

*Ductility* is the ability of a material to deform without breaking. Aircraft are dynamic machines. They must be able to change their shape under loads and to do this they must be ductile. The ductility of a material is measured as a percentage of its original size.

### Brittleness

*Brittle* is the opposite of ductile. Brittle materials do not change shape readily under applied loads. They break with little or no warning. Plate glass is a good example of a brittle material. Aircraft materials cannot be too brittle, they must be able to change their shape when loads are applied.

## LOADS AND STRESSES (More illustrations in the Appendix)

When a load is applied to a metallic structural material, the metallic bonding resists a change in the atomic arrangement. This resistance is called *stress.* Stress is measured in pounds per square inch, psi, in the U.S. customary system of measurements. Stress refers to a definite plane passing through a given point on the structural member.

This book was written with the user of structural materials, rather than the designer of them, in mind. Therefore the maximum stresses developed in a material, those that may lead to structural failures, are of prime interest.

One simplifying assumption that is made in this discussion is that tension (pulling), compression (pushing), and shearing (slicing) loads are uniformly distributed and that the resulting stresses are also uniformly distributed over the area of the structural member.
Stresses can be classified as those that are perpendicular, parallel, or oblique to the applied load.

### Normal Loads and Stresses

Consider a structure that has forces, P, applied that tend to pull against each other. These are called *tension forces* and the resulting stress is called *tension stress,* $f_t$.
The tension stress acts on a plane perpendicular (normal) to the load. The value of the stress is simply the load, P, divided by the cross sectional area, A.

$$f_t = \frac{P}{A} \tag{5.1}$$

Engineers have devised a system to help them visualize stresses. They mentally cut the member and take the cut member aside and examine the stresses needed to place the member in equilibrium. This is called "taking a free body." Such a procedure is shown in Figure 5.1. It shows that the stresses resist the tension load on the member.

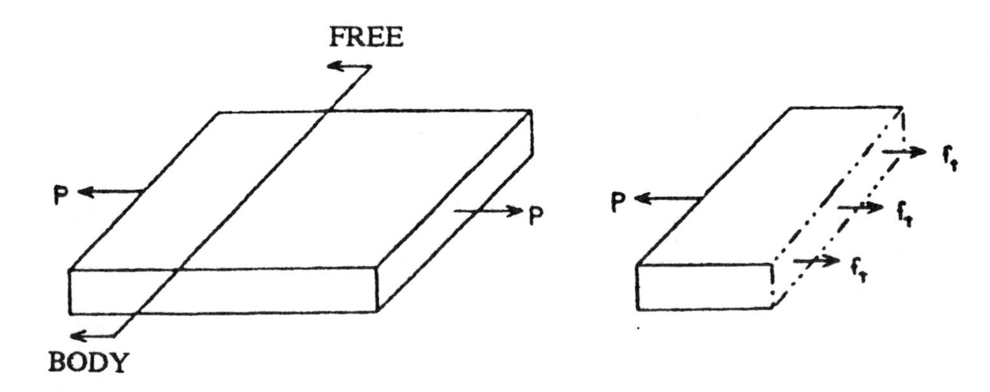


*Figure 5.1 Normal tension loading and stress.*

Consider that the forces acting on a structure are pushing against each other. These are called *compression forces.* The axial stress is *compression stress,* $f_c$. This loading acts as shown in Figure 5.2.

The compression stresses act in the opposite direction to the tension stresses. They act on a plane that is normal to the load. Compression stress is calculated in the same way as tension stress, $f_c = P/A$.

It should be noted that most compression failures in aircraft structures do not occur because of compression overstress. They fail as columns by buckling. This will be discussed later.

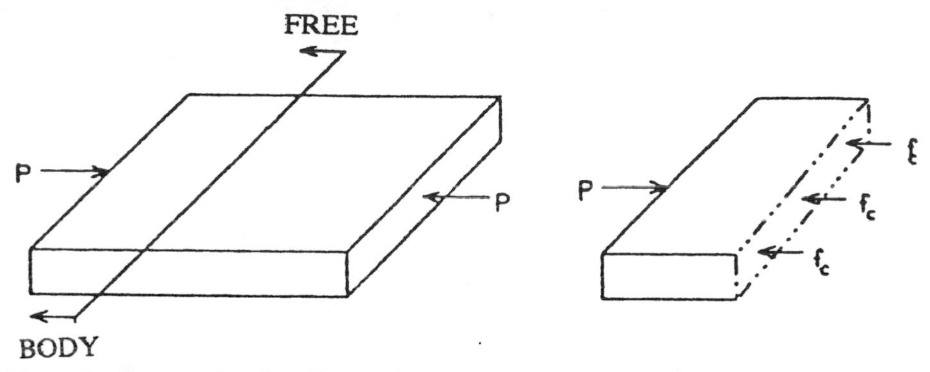


*Figure 5.2 Normal compression loading and stress.*

## Shear Loads and Stresses

Loads that are acting parallel to the plane of the structure are called shear loads and the resulting stress is shear stress, $f_s$. In shear loading the loads can be acting toward each other or away from each other. But unlike tension or compression loads, they do not act in line with each other. The loads pass by each other just like the blades of a pair of scissors.

Figure 5.3(a) shows shearing loads. Shear stresses are shown in Figure 5.3(b). They are calculated as in equation 5.1, $f_s = P/A$. If the material fails due to high shear stress, the plane of failure is not perpendicular to the applied loads, but is parallel to and between them.

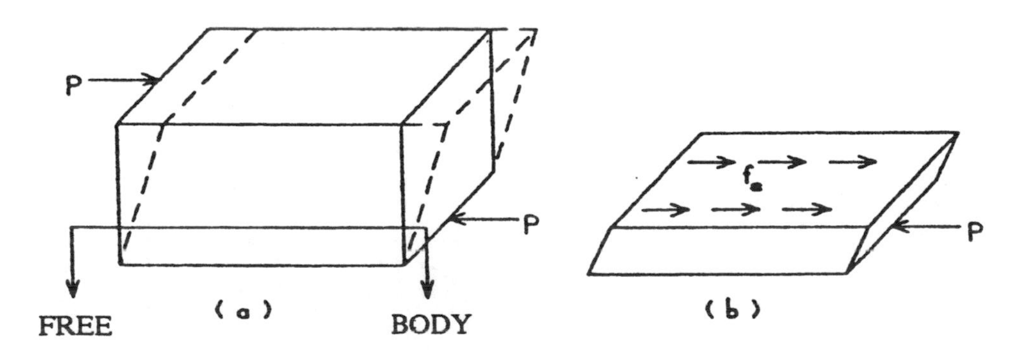


*Figure 5.3 (a)Shear loading and (b)Shear stress.*

## STRESSES ON OBLIQUE SECTIONS

The stresses discussed so far are called primary stresses. They act on planes that are perpendicular to the loads (tension and compression) or parallel to the loads (shear).

There are other stresses that are acting on the material simultaneously. These do not act on the same planes as the primary stresses. These are called component stresses and act on planes that are oblique to the applied loads.

### Tension and Compression Loads

Consider a material that is in tension as shown in Figure 5.4(a). For equilibrium conditions to exist the load P on the top of the member must be resisted by an equal and opposite force on the bottom. If a plane is cut through the member perpendicular to the applied load, (plane A-A in Fig.5.4(b)), only tension stresses act on this plane.

Next consider the member to be cut along plane B-B in Figure 5.4(c). Half of the load acts on each side of B-B and no stresses are developed along the cut surface.

Assume that the member has a thickness of one inch and that the member is cut along an oblique plane C-C as shown in Figure 5.4(d). The force P, at the top, must be resisted by an equal and opposite force P, on the cut surface. As we are interested in the loads that exist on the cut surface, we shall resolve the lower force into two components. The component normal to the cut surface is a tension load, $P_t$ and the component that is parallel to the surface is a shear load, $P_s$.

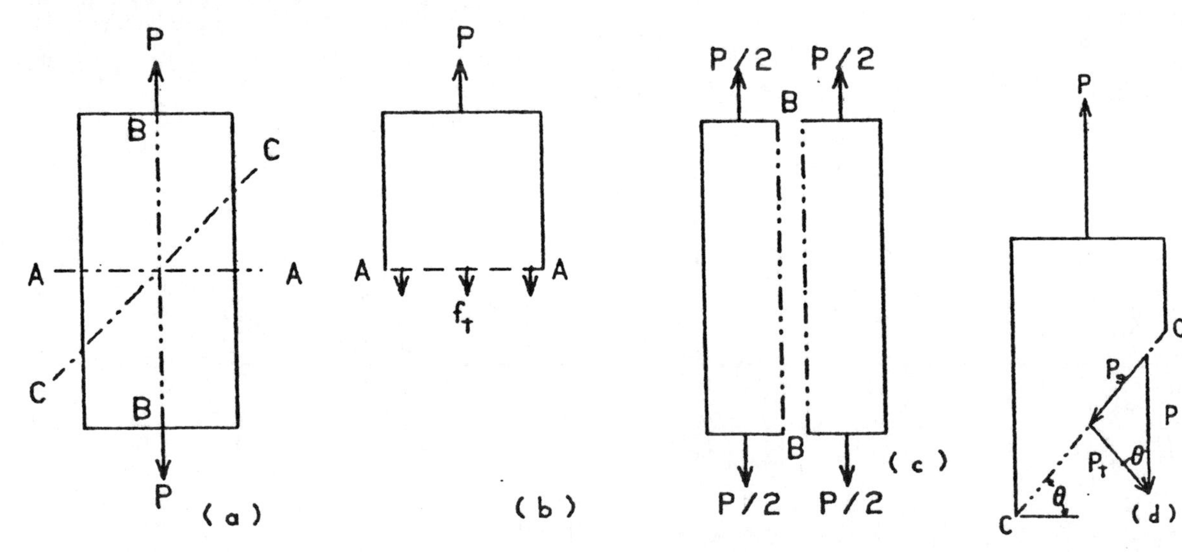


*Figure 5.4 Component loads of a tension load.*

To find the shear stress on the cut surface we divide the shear load by the shear area. The shear stress, $P_s$, in terms of the perpendicular stress, P, is:

$$P_s = P \sin \theta$$

The area of the cut surface, $A_s$, in terms of the perpendicular area, A, is:

$$A_s = \frac{A}{\cos \theta}$$

Thus

$$f_s = \frac{P_s}{A_s} = \frac{P(\sin\theta\cos\theta)}{A} = f_t(\sin\theta\cos\theta) \tag{5.2}$$

Analysis of Equation(5.2) shows that fs will be a maximum when θ is 45°. When θ is 45°, (sinθcosθ) = .5. Thus the maximum shear stress is equal to one-half of the maximum tensile stress.

$$f_{s(max))} = \frac{1}{2} f_{t(max)}$$

Compressive loading follows the same argument and produces shear stress as well as compressive stress. The values are the same as for tension. Maximum shear stress occurs on the 45° plane and the value of the maximum shear stress is one-half of the maximum compressive stress.

The important point in this discussion is that, while a member is undergoing tension or compression loading, there are shear stresses being developed in the member as well as the direct tension or compression stresses.

Many materials such as aluminum, magnesium and other ductile materials have somewhat low shear strength, as compared to their tensile strength. These materials will usually fail in shear on a 45° plane to the applied tension or compression load. Brittle materials, on the other hand, have high shear strength compared to their tensile strength. They will fail in tension, not in shear as shown in Figure 5.5.

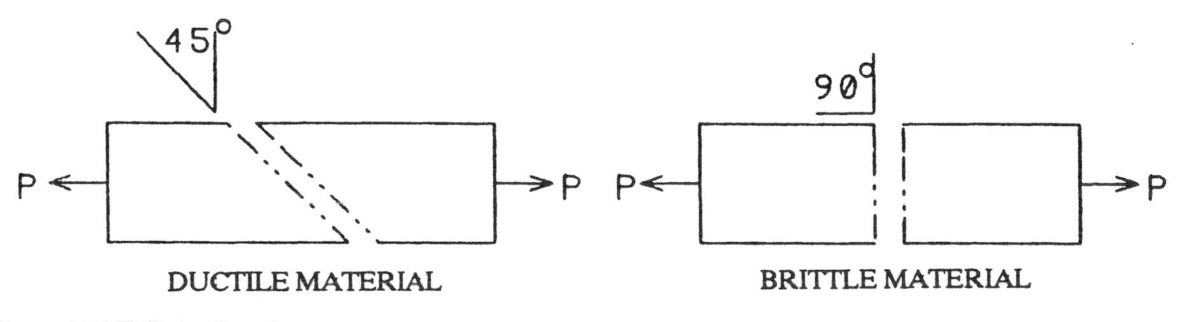


***Figure 5.5 Failure directions.***

## SHEAR LOADS

To consider the stresses on oblique sections of members that are subjected to shear loads, consider a large flat plate, as shown in Figure 5.6. The applied shear loads are shown by the large arrows at the sides of the plate. These shear loads produce shear stresses in the plate that are transmitted uniformly through the material.

The small square in the center of the plate represents the "free body." We will examine this to find the shear stresses and the stresses on the oblique sections. The applied shear loads at the sides of the plate are transmitted to the sides of the square and are shown by the small arrows.

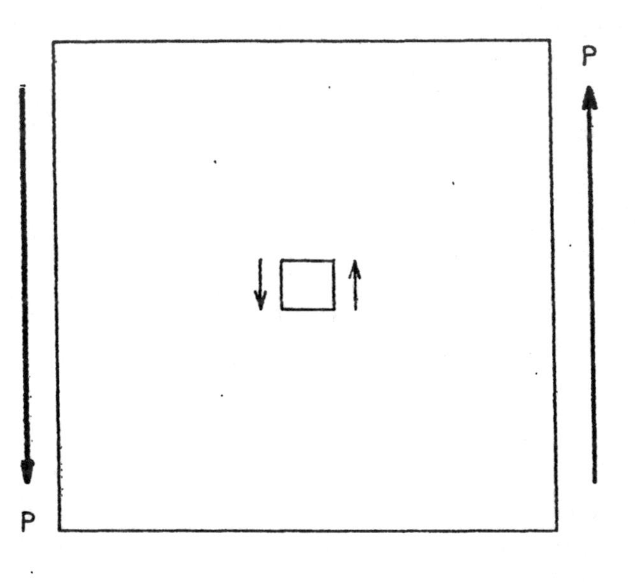


***Figure 5.6 Shear loads on a flat plate.***

The square in Figure 5.6 is removed, enlarged for clarity, and shown in Figure 5.7(a). The shear loads at the side of the square produce a counter-clockwise rotating moment on the square. We know, from observation, that the square does not rotate. We can conclude that the atoms at the top and bottom of the square resist being displaced. They produce resisting forces as shown by the arrows at the top and bottom of the square shown in Figure 5.7(a).

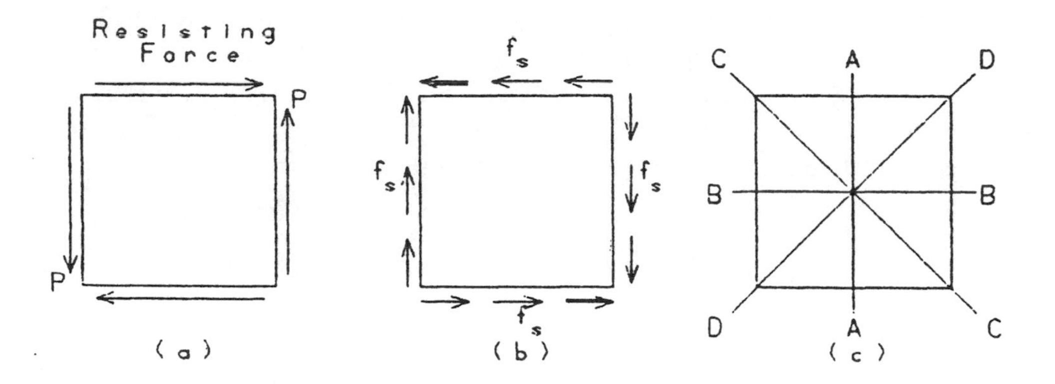


*Figure 5.7 (a)Forces; (b)Stresses; (c)Sections of investigation on a square.*

Figure 5.7(b) shows the shear stresses resulting from the shear loads and the resisting forces.

Figure 5.7(c) shows the various sections that we will now investigate. First we take "free bodies" along section lines A-A and B-B, as shown in Figure 5.8.

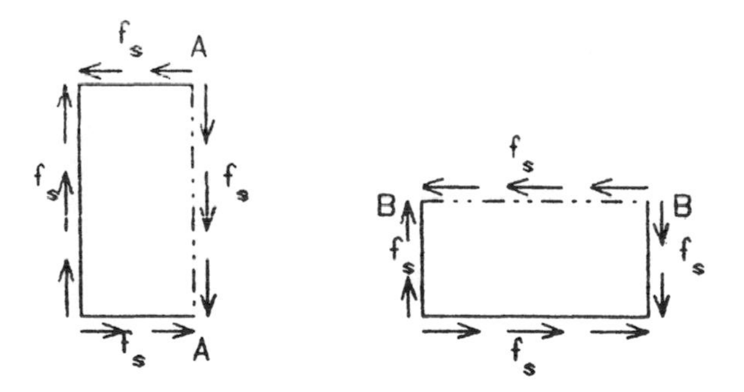


*Figure 5.8 Stresses on Sections A-A and B-B.*

For equilibrium to exist it shows that only shear stress occurs on section A-A and B-B. Theoretically the shear stress is the same on both sections, but practically most shear failures will occur on section A-A (parallel to the applied load).

Figure 5.9(a) shows the stresses existing on section C-C. Taking an imaginary cut along C-C and removing the free body (lower left half of the square) shows the shear stresses acting on the section. To find the stresses on the cut surface, the shear stresses are resolved into components parallel and perpendicular to the surface. The stresses that are parallel to the surface are compression stresses and cancel each other, but those that are perpendicular to the surface are tension stresses and are cumulative. They exert tension stress on the surface.

It shows that no shear or net compression stress exists on this surface.

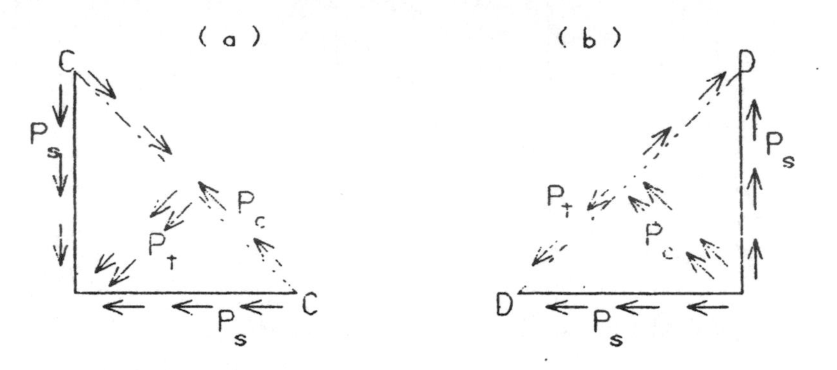


***Figure 5.9 Stresses on: (a) section C-C; (b) section D-D.***

Figure 5.9(b) shows the stresses existing on section D-D. We take the lower right half of the square as a free body and resolve the shear stresses into components parallel and perpendicular to the surface. The tension stress components that are parallel to the surface cancel, but the stress components that are perpendicular to it are cumulative. These produce compression stresses against the surface.
Thus only compression stresses occur on section D-D.

There are two load components acting on both section C-C and section D-D surfaces. This means that the resulting tension stress and compression stress are equal to the shear stress.

$$f_{t(max)} = f_{c(max)} = f_{s(max)} \quad (5.4)$$

All stresses have equal values, act simultaneously, but on different planes.

To recap, applied shear loads produce shear stresses that are a maximum on sections that are parallel or perpendicular to the direction of the shear loads.

They also produce tension and compression stresses that are maximum on sections that are at 45° angles to the applied shear loads.

The maximum values of all stresses are equal.

## ENGLISH SYMBOLS

A = area of material, sq.in.
f = compression stress, psi.
$f_s$ = shear stress
$f_t$ = tension stress
P = applied load, lb.

## GREEK SYMBOL

θ (theta) = angle of cut surface, degrees

## EQUATIONS

5.1 $f_t = \frac{P}{A}$

5.2 $f_s = f_t(\sin\theta\cos\theta)$

When a tension load is applied:

5.3 $f_{s(max)} = \frac{1}{2} f_{t(max)}$ *($f_{s(max)}$ is on 45°)*

When a shear load is applied:

5.4 $f_{t(max)} = f_{c(max)} = f_{s(max)}$ *($f_{t(max)}$ and $f_{c(max)}$ are on 45°)*

## REVIEW PROBLEMS

1. A metal bar in pure tension will have ______ stress on a plane parallel to the applied load.
   (a) tension (b) shear (c) zero

2. A pure tension load produces a primary stress of tension. There is also a component stress that is:
   (a) shear stress and is equal to the tension stress;
   (b) compression stress and is equal to the tension stress;
   (c) shear stress and is equal to 1/2 of the tension stress.

3. The component stress in problem 2 above acts:
   (a) on a plane that is parallel to the tension stress
   (b) on a plane that is 45 degrees to the tension stress
   (c) on a plane that is 9 degrees to the tension stress.

4. The formula, f = P/A is used to calculate stress resulting from three kinds of loading. These are:
   (a) shear, tension and bending
   (b) compression, shear and tension
   (c) shear, torsion and bending.

5. Assume that shear loads are being applied to a thin aluminum panel such as shown in Figure 5.6. Failure will be:
   (a) shear and will fail on a plane parallel to the large arrows
   (b) compression buckling from upper left to lower right
   (c) tension tearing from upper left to lower right
   (d) compression buckling from upper right to lower left.

6. If you are examining an aluminum skin of a light airplane wreck and find 45 degree fractures of the edge of the skin, you can conclude that the load that caused the break was:
   (a) tension
   (b) compression
   (c) shear.

7. Find $f_{T\,MAX}$ and $f_{S\,MAX}$, compare these stresses with ultimate tension stress of 64,000 psi and ultimate shear stress of 30,000 psi. Show failure direction on sketch.

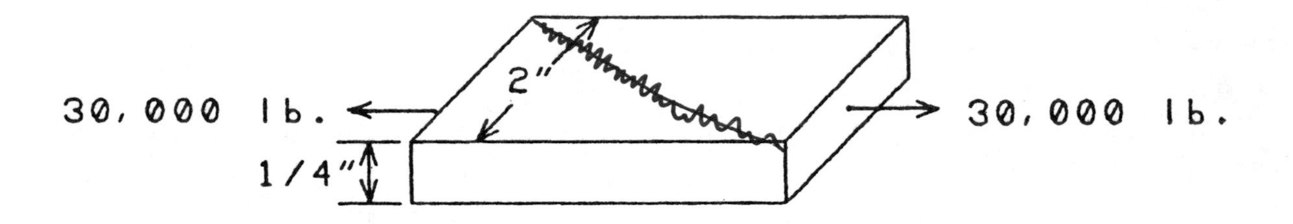


8. Same as problem 7 except ultimate tensile stress is 90,000 psi. and ultimate shear stress is 60,000 psi. Show failure direction on sketch.

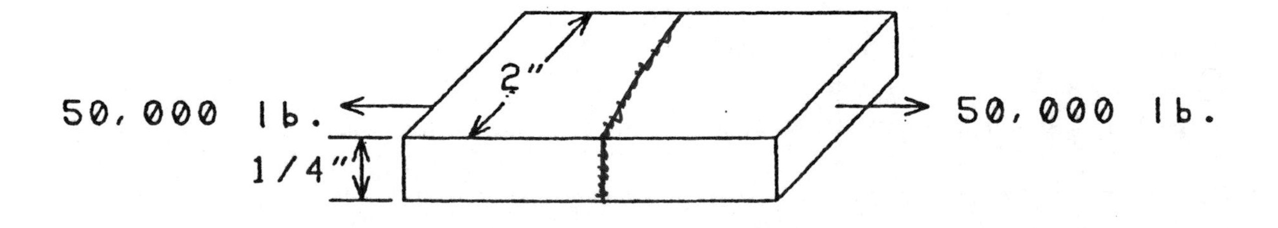

# CHAPTER SIX

## STRESSES FROM TORSION AND BENDING LOADS

So far we have discussed the simple cases of pure tension, compression and shear forces producing the corresponding tension, compression and shear stresses.

There are other types of loads, but only the three stresses discussed are possible.

### STRESSES FROM TORSION LOADS (More illustrations in the Appendix)

When we think of torsion carrying members in aircraft, we usually think of drive shafts in the engine section. But, the wings and fuselage are torque carrying structures also.

The analysis of a drive shaft is much simpler, so this discussion will be limited to a drive shaft.

Consider a drive shaft with equal and opposite torque forces being applied to the ends. Picture a square drawn on the surface of the shaft. The amount of torque on the shaft is the same at all points along the shaft between the two torque forces. Thus we can show the torque forces acting at the sides of the square.

This is shown in Figure 6.1.

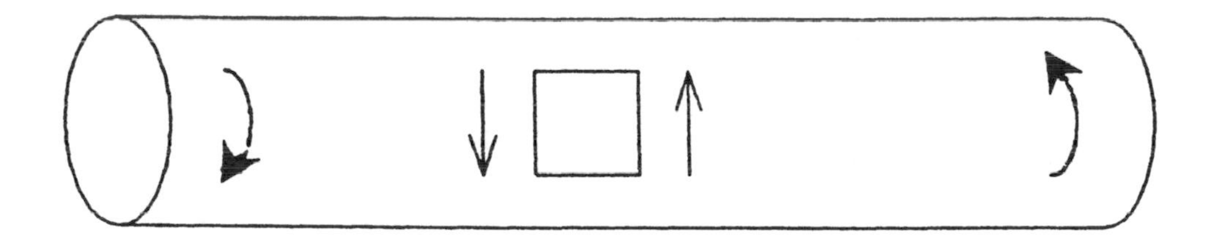

***Figure 6.1 Drive shaft under torque forces.***

These two forces are shearing forces, similar to the pure shear forces discussed in Chapter Five. The analysis of the stresses acting on the square is the same as those that were discussed on pages 44-46.

Figure 6.2 shows the square removed as a free body and the maximum stress planes. From our earlier discussion we found that the maximum stresses were equal in value, $f_{T\,MAX} = f_{C\,MAX} = f_{S\,MAX}$, but they acted on different sections.

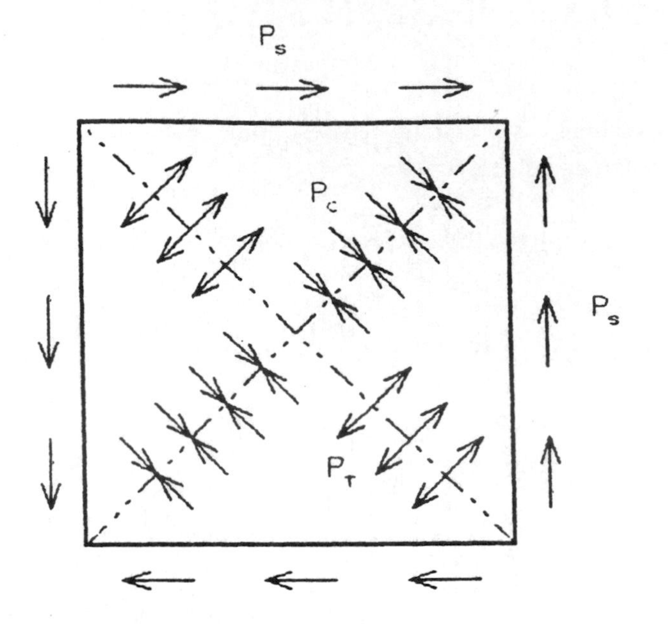


*Figure 6.2 Stresses on surface of torqued drive shaft.*

A drive shaft has a *neutral axis* where shear stresses are zero. It is at the center of the shaft. As we examine shear stresses at other locations we find that they increase linearly as the distance from the neutral axis, NA. They are a maximum on the outermost fiber, where the distance from the NA is equal to r, the radius of the shaft. The shear stress distribution is shown in Figure 6.3.

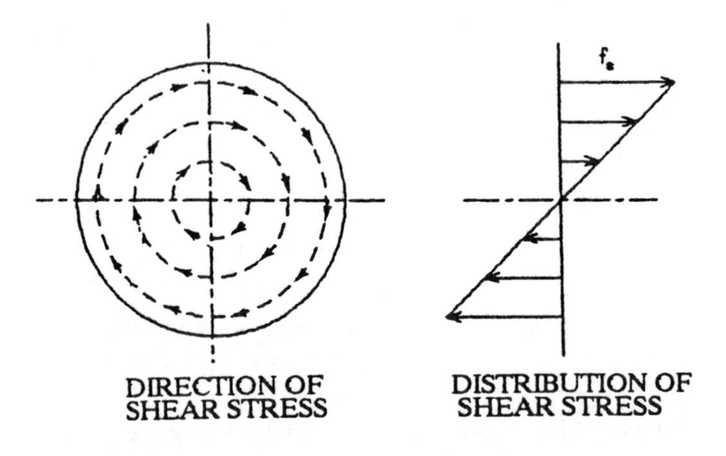


*Figure 6.3 Shear stress distribution in drive shaft.*

Besides the values of torque, T, and radius, r, we need to know one other factor to calculate the maximum shear stress. It is called the *polar moment of inertia*, J, and represents the size and shape off the cross section of the shaft. Equations for J for solid and hollow shafts are:

$$J = \frac{1}{32}\pi(D_o^4) \qquad \textit{(solid circular shaft)} \tag{6.1}$$

$$J = \frac{1}{32}\pi(D_o^4 - D_i^4) \qquad \textit{(hollow circular shaft)} \tag{6.2}$$

Where J = polar moment of inertia, $in^4$.
$D_o$ = outside diameter, in.
$D_i$ = inside diameter, in.

## Calculation of Stress from Torsion Loads

Maximum shear stress is calculated by:

$$f_{s(max)} = \frac{Tr}{J} \tag{6.3}$$

Where $f_{s(max)}$ = maximum shear stress, psi.
T = torque, in-lb.
r = outside radius, in.
J = polar moment of inertia, $in^4$.

In bending we saw the advantage of locating the material away from the neutral axis in reducing stress levels. Similar advantages occur in torqued shafts. To illustrate this, two example problems are presented.

**Example Problem 1.**

a. Calculate the area of the pictured shaft.
b. Calculate the maximum shear stress.

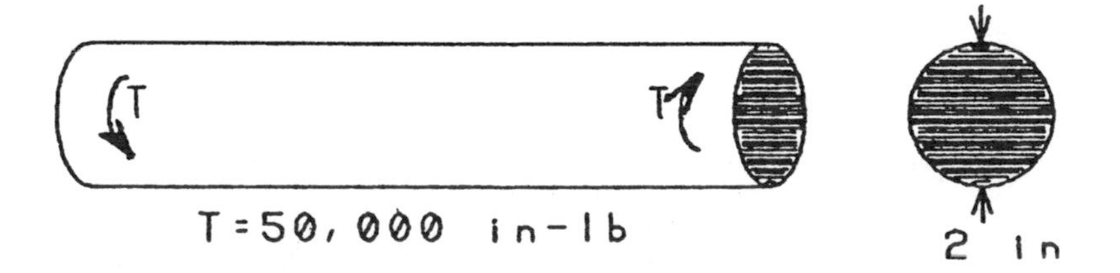


Solution a.: Area = $\pi r^2$ = $\pi$ sq.in.
Solution b.: T = 50,000 in-lb., r = 1 in., J = $\pi(2^4)/32$ = $\pi/2$ $in^4$.
$f_{S\,MAX}$ = 100,000/$\pi$ = 31,831 psi.

**Example Problem 2.**

a. Calculate the area of the pictured shaft.
b. Calculate the maximum shear stress.

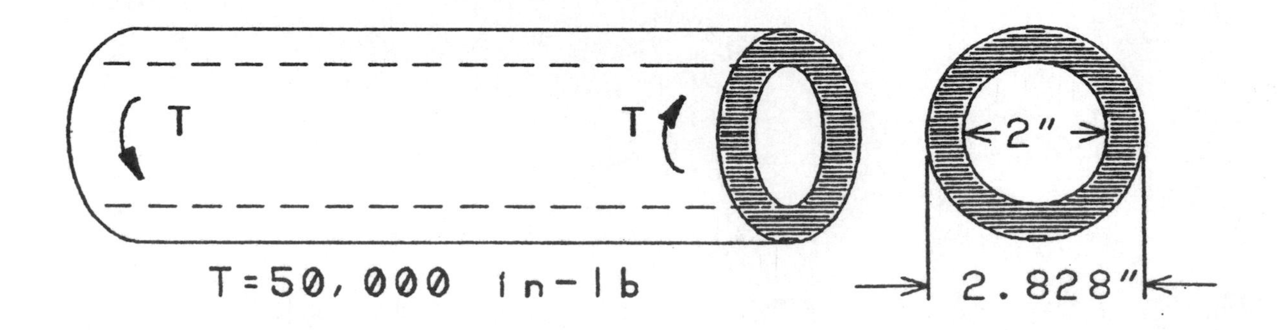


Solution a.: Area = $\pi r_o^2 - \pi r_i^2 = 2\pi - \pi = \pi$ sq.in.
Solution b.: T = 50,000 in-lb.
$r = 1.414$ in.
$J = \pi(2.828^4 - 2^4)/32 = 4.7124$ in$^4$.

$f_{S\,MAX} = (50{,}000)(1.414)/4.7124 = 15{,}003$ psi.

Here we see the advantage of locating the material away from the neutral axis. Each shaft in the above examples has the same area (same weight), but the hollow shaft has less than half the stress of the solid shaft.

## SHAFT FAILURES (More illustrations in the Appendix)

Tension, compression and shear stresses are all equal in a drive shaft subjected to torsion loads. So the type of failure to be expected depends upon the shaft material (ductile or brittle) and the type of shaft (solid or hollow). We will analyze each possibility separately.

### Hollow Shaft-Ductile Material

This is typical of helicopter anti-torque rotor drive shafts. The hollow shaft allows room for the metal to buckle. Solid shafts do not allow this type of failure. The ductile material allows the metal to deform, thus buckling is possible. Brittle materials cannot deform so this type of failure is not possible for them.

The hollow ductile material shaft will fail by buckling as shown in Figure 6.4

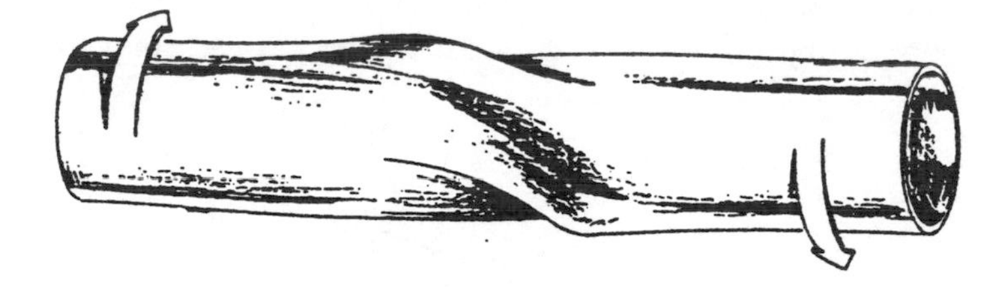

***Figure 6.4 Buckling failure of hollow shaft of ductile material.***

Note that a line drawn between the points of the torque arrows shows the line of the buckles. Thus the torque direction can be determined if this kind of buckling occurs.

## Solid Shaft-Brittle Material

This type of shaft is found in engine accessory drive shafts. Brittle materials are strong in compression and shear, but somewhat weak in tension. Thus we expect to see this type of shaft fail in tension. It does this by a pulling itself apart on a helical plane, 45° to the axis of the shaft. This is shown in Figure 6.5.

To understand this type of break, apply torsional forces to the ends of a piece of blackboard chalk and observe the failure.

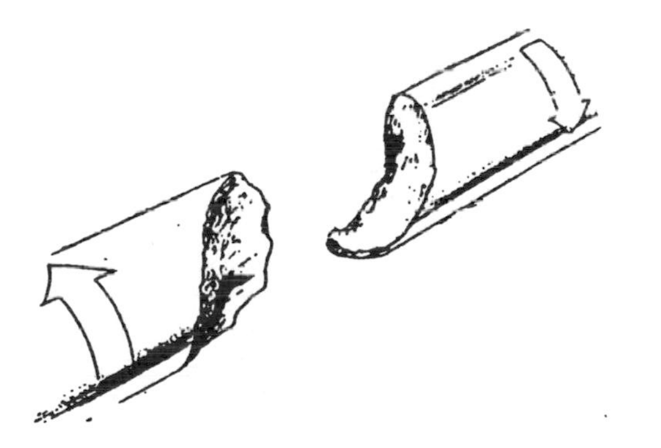

***Figure 6.5 Tension failure of a brittle solid shaft under torsion loading.***

## Solid Shaft-Ductile Material

Ductile materials are relatively weak in shear as we discussed in Chapter Five. This type of shaft then fails in shear on a plane 90° to the axis of the shaft. This is shown in Figure 6.6.

## *Stresses From Torsion and Bending Loads*

If the shaft is short the tendency of a shear failure increases. Close observation of the sheared surface reveals that the fracture starts at the outside fibers, where stresses are a maximum, and spiral inward toward the center.

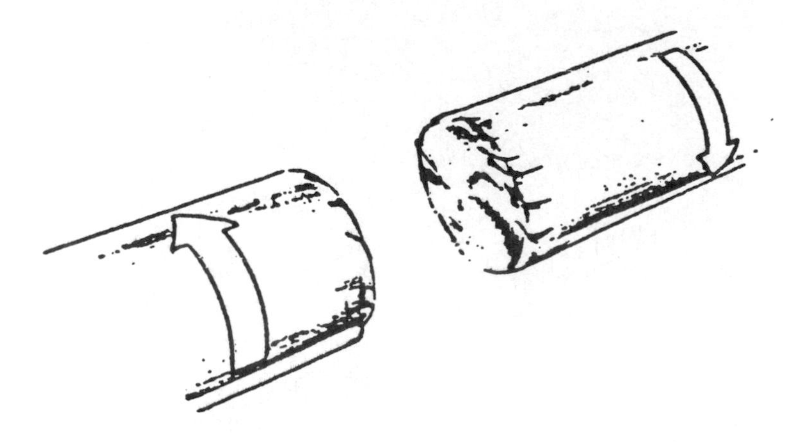

*Figure 6.6 Shear failure of a solid shaft of ductile material.*

## Hollow Shaft-Brittle Material

Only rarely will a hollow shaft of brittle material be used in aircraft parts. One such use is in hollow steel propeller blades. These carry some torsion loads. The brittle materials cannot deform, so buckling is not possible even though the shaft is hollow. As the shear strength is greater than the tension strength, tension failures on the 45° plane will result. This is similar to the solid shaft failures of brittle materials.

## BENDING LOADS(More illustrations in the Appendix)

Next, we will discuss bending of beams. The beams that we consider are bent by forces acting perpendicular to the axis of the member. Simple cases of beam bending are (1) the cantilever beam (Figure 6.7(a)) and (2) the simple or end supported beam (Figure 6.7(b)).

The cantilever beam is one that is fixed at one end and free at the other. It is similar to a diving spring board or an airplane wing. The end supported beam is one that rests between two supports.

The simple beam with one load acting at its center is shown in Figure 6.7(b). This type of beam is roughly similar to the main spar of an airplane. The weight of the plane is considered to be the central load and the main landing gears are the end supports.

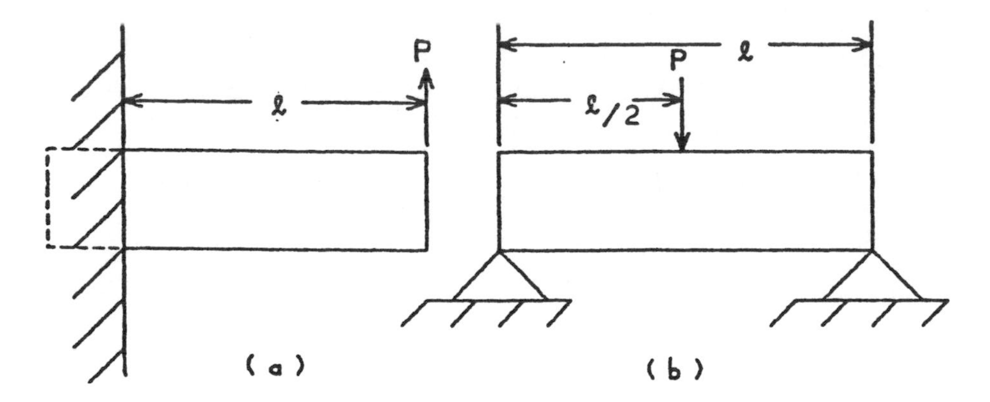


*Figure 6.7 (a) Cantilever beam; (b) Simple beam.*

Before we can analyze either of the above beams, we must understand a basic fact. The beam with its applied load(s) is in an equilibrium condition. The conditions that must be satisfied for equilibrium are:

Algebraic sum of horizontal forces = 0
Algebraic sum of vertical forces = 0
Algebraic sum of moments = 0

## Equilibrium Forces and Moments

Consider the cantilever beam shown in Figure 6.7(a). To satisfy the condition of equilibrium of forces there must be a downward force exerted on the beam by the wall support equal to the load P. The load P also develops a counter-clockwise moment M about the wall support.

$$M = Pl \tag{6.4}$$

For equilibrium of moments, this moment must be counteracted by an equal and opposite moment in the beam at the wall. The beam is shown in equilibrium in Figure 6.8.

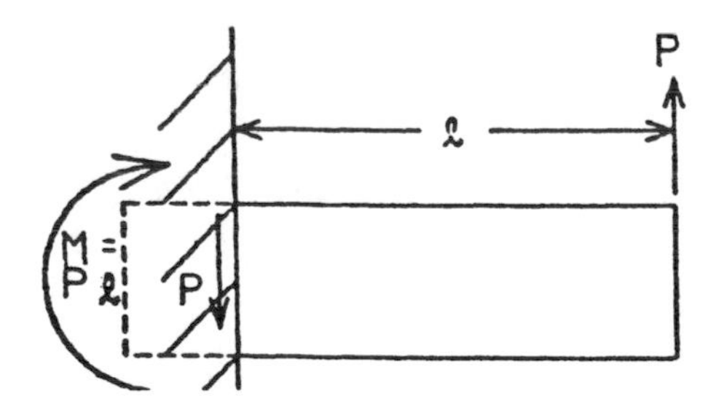


*Figure 6.8 Cantilever beam in equilibrium.*

Consider the simple beam shown in Figure 6.7(b). To satisfy the condition of equilibrium of forces there must be a reaction at each end supports equal to P/2. At first glance the beam seems to be in equilibrium because the sum of the moments are equal to zero.

But there are internal stresses developed in the beam and these create internal moments. To analyze these we mentally cut the beam in the center and take either side as a free body. In Figure 6.9 we have taken the left side as a free body and find that the internal moment at the center of the beam is:

$$M = \left(\frac{P}{2}\right)\left(\frac{l}{2}\right) = \frac{Pl}{4} \qquad (6.5)$$

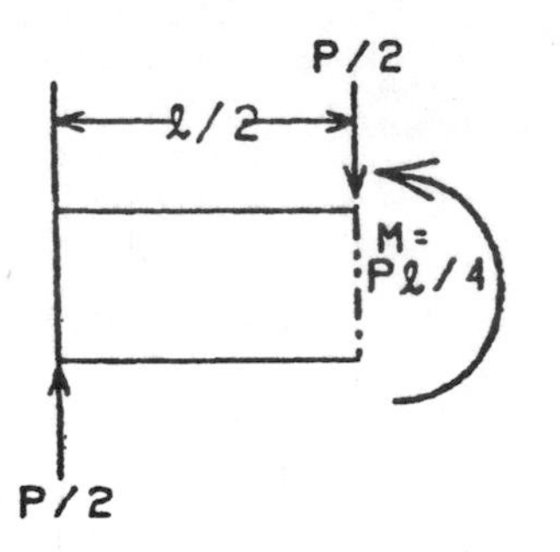


*Figure 6.9 Simple beam in equilibrium.*

## Stresses in Loaded Beam

Consider the simple beam with a single downward load applied in the center. The beam will deflect as shown in Figure 6.10(a). The deflection is exaggerated for clarity in the drawing. Vertical equally spaced lines drawn before the load is applied now become radials. The distance between the lines at the top of the beam, (GB), is shortened to (GB'), indicating that the fibers are in compression.

The opposite condition exists on the bottom of the beam. The distance between the lines (CD) is increased to (CD') and the fibers are in tension. At some point between the top and bottom of the beam the distance (EF) is not changed. This point lies on the neutral axis (N-A).

The neutral axis passes through the *centroid* (center of area) of the beam cross section. There is no tension or compression stress along the neutral axis. The stress increases linearly as the distance (y) from the neutral axis increases.

Maximum tension stresses occur at the outermost fibers on the surface farthest away from the applied load. Maximum compression stress occurs on the outside fibers on the opposite side.
Stress distribution is shown in Figure 6.10(b).

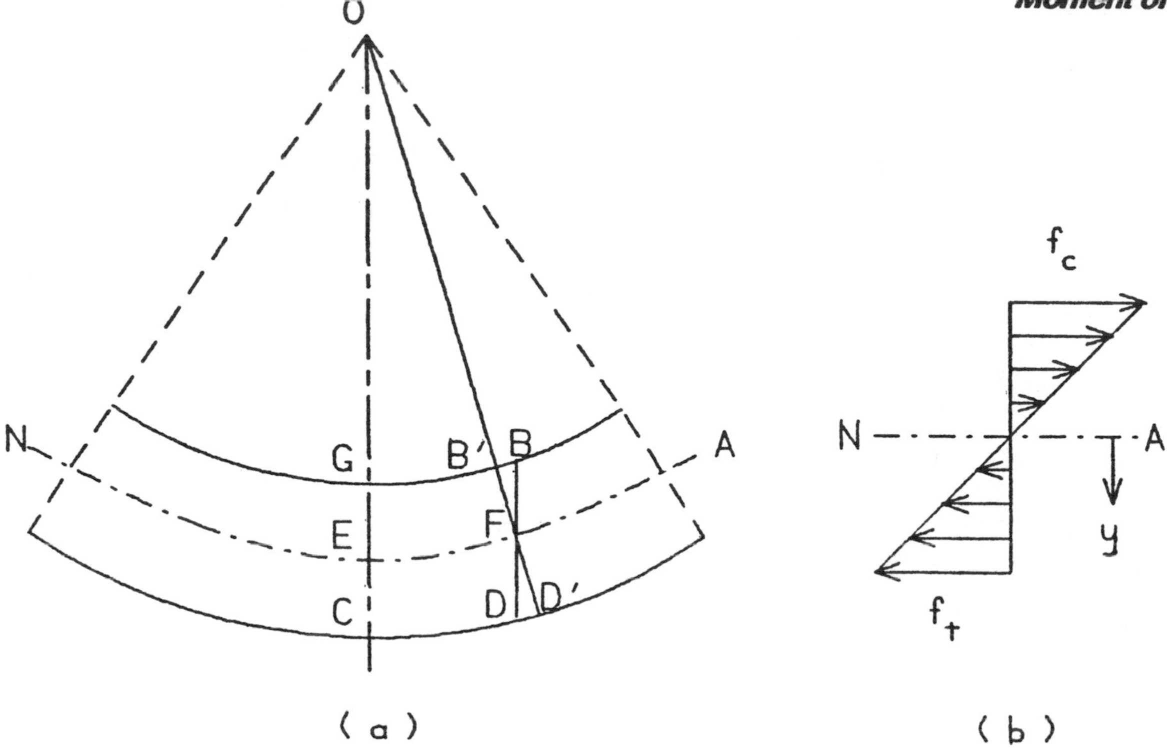


*Figure 6.10 Simple beam (a) Deflection; (b) Stresses on beam.*

## MOMENT OF INERTIA

One other factor enters the calculation of primary stresses developed by bending. This is called the *moment of inertia*, I. The words "moment" and "inertia" in the phrase moment of inertia have different meanings than if the same words are used separately. The moment of inertia of the cross sectional area of a beam with respect to the neutral axis is: "the sum of the products of each elementary area of the cross section multiplied by the square of its distance from the neutral axis." Engineering handbooks list formulas for calculating I for many structural sections. A few of these are listed here:

$$I = \frac{1}{12}(bh^3) \quad (rectangular) \qquad I = \frac{1}{64}(D_o^4)(solid\ circular)$$

$$I = \frac{1}{64}(D_o^4 - D_i^4) \quad (thick\ wall\ tube) \qquad I = \frac{1}{8}(D_o^3 t) \quad (thin\ wall\ tube)$$

Where I = moment of inertia about neutral axis, $in^4$.
b = rectangle base width, in.
h = rectangle height, in.
$D_o$ = outside diameter, in.
$D_i$ = inside diameter, in.
t = tube wall thickness, in.

## Calculation of Stress from Bending Loads

Both tension and Compression stresses are calculated by:

$$f_t \text{ or } f_c = \frac{My}{I} \tag{6.6}$$

Where M = bending moment of beam, in-lb.
y = distance from neutral axis, in.
I = moment of inertia, $in^4$.

There are no tension or compression stresses at the neutral axis as y = 0. As the distance from the neutral axis, y, increases, these stresses generate and increase. If the material of the beam is placed where stress is a maximum, the material will be used more efficiently and weight will be reduced. Moving the material away from the neutral axis increases the moment of inertia. Because "I" is in the denominator of equation (6.6), high values of it will reduce the stress values.

**Example Problem 1.** Calculate $f_{t(max)}$ of this beam.

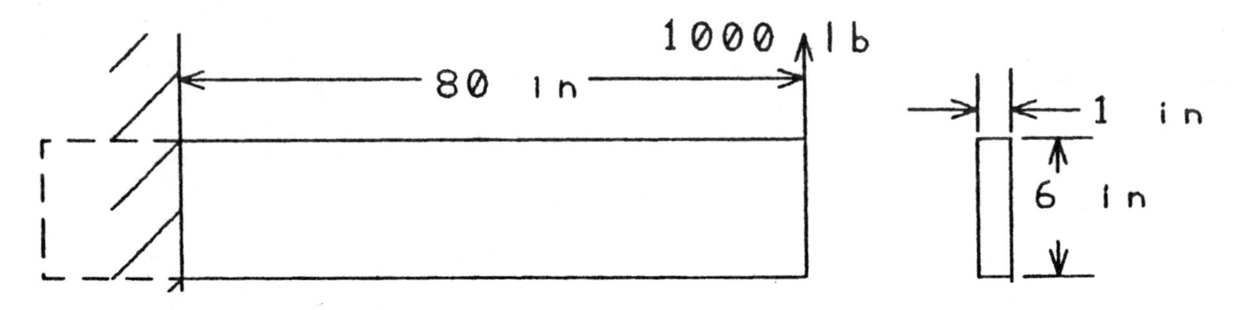


Solution: M = 80,000 in-lb., y = 3 in., b = 1 in., h = 6 in.
$I = (1)(6^3)/12 = 18\ in^4$.
$f_{t(max)} = My/I = (80{,}000 \times 3)/18 = 13{,}333$ psi.

**Example Problem 2.** Calculate $f_{t(max)}$ of this beam.

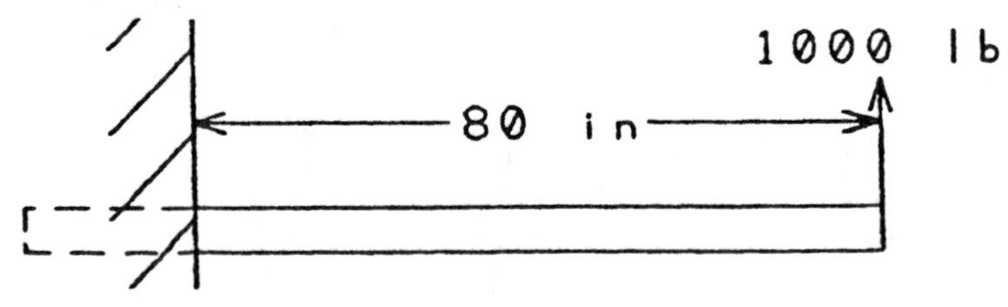


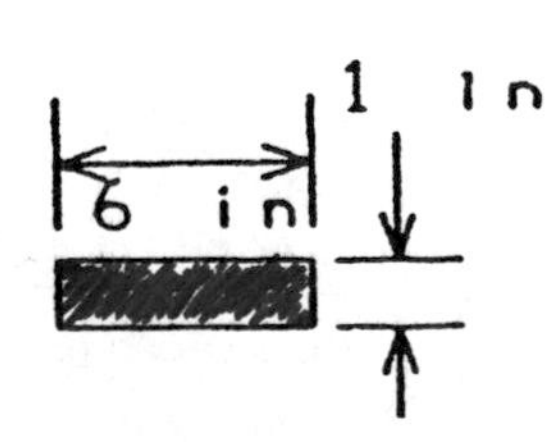


Solution: M = 80,000 in-lb.,y = 0.5 in., b = 6 in., h = 1 in.
$I = (6)(1^3)/12 = 0.5\ in^4$.
$f_{t(max)} = My/I = (80{,}000 \times 0.5)/0.5 = 80{,}000$ psi.

Comparing the two solutions,it is seen that both beams have the same cross sectional area (and thus the same weight). But the first beam, which has the material located away from the neutral axis, produces lower stress levels. Here, it has only 16.7 percent of the stress of the second beam.

This explains why engineers use I-beams, T-beams, and other shapes that move the material away from the neutral axis.

A 12 inch high I-beam has a moment of inertia of 319.3 $in^4$. A rectangular beam with the same height and area (and thus the same weight) has a moment of inertia of only 192.5 $in^4$. The rectangular beam could only support 60.3 percent of the load of the I-beam. To carry the same load the I-beam would weigh less than half as much as the rectangular beam.

In addition to the tension and compression stresses in a beam undergoing bending loads, there are also shear loads generated in the beam. The shear stresses are not uniform over the area. The average shear stress is the vertical shear load, V, divided by the cross section area, A. $f_{s(av)} = V/A$.

Shear stress in beams is important mainly in timber beams, concrete beams, and some built-up steel beams. As we are interested in airplanes, this subject will not be pursued further.

**SUMMARY OF LOADS AND STRESSES**

| LOAD | PRIMARY STRESS 90° OR 0° | FORMULA | COMPONENT STRESS (45°) | FORMULA |
|---|---|---|---|---|
| TENSION | TENSION | $f = P/A$ | SHEAR | $f_s = 1/2\ f_t$ |
| COMPRESSION | COMPRESSION | $f = P/A$ | SHEAR | $f_s = 1/2\ f_c$ |
| SHEAR | SHEAR | $f = P/A$ | TENSION & COMPRESSION | $f_t = f_c = f_s$ |
| BENDING | TENSION & COMPRESSION | $f = My/I$ | SHEAR | $f_{av} = V/A$ |
| TORSION | SHEAR | $f = Tr/J$ | TENSION & COMPRESSION | $f_t = f_c = f_s$ |

*Stresses From Torsion and Bending Loads*

## SYMBOLS

I = moment of inertia, $in^4$.
J = polar moment of inertia, $in^4$.
M = bending moment, in-lb or ft-lb.
NA = neutral axis position
V = vertical shear load, lb.
y = distance from NA, in.

## EQUATIONS

6.1 $J = \frac{1}{32}\pi(D_o^4)$ *(solid circular shaft)*

6.2 $J = \frac{1}{32}\pi(D_o^4 - D_i^4)$ *(hollow circular shaft)*

6.3 $f_{s(max)} = \frac{Tr}{J}$

6.4 $M = Pl$

6.5 $M = \left(\frac{P}{2}\right)\left(\frac{l}{2}\right) = \frac{Pl}{4}$

6.6 $f_t \text{ or } f_c = \frac{My}{I}$

## REVIEW PROBLEMS

1. The principal stress resulting from torsional loading is:
   (a) tension; (b) compression; (c) shear; (d) torsion.

2. The component (secondary) stresses resulting from torsional loading are:
   (a) shear and tension; (b) tension and compression; (c) torsion and bending.

3. Calculate $f_{t(max)}$ for this beam.

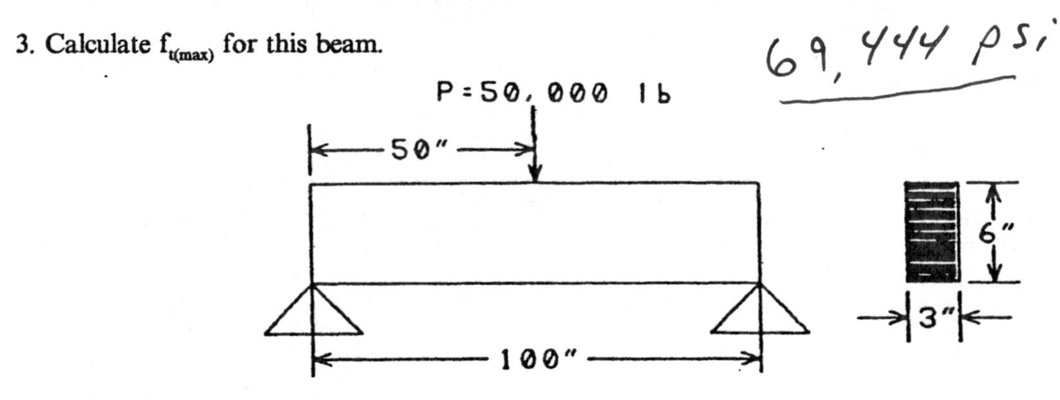


4. Calculate maximum $f_t$, $f_c$, and $f_s$. Show failure direction for this solid brittle shaft.

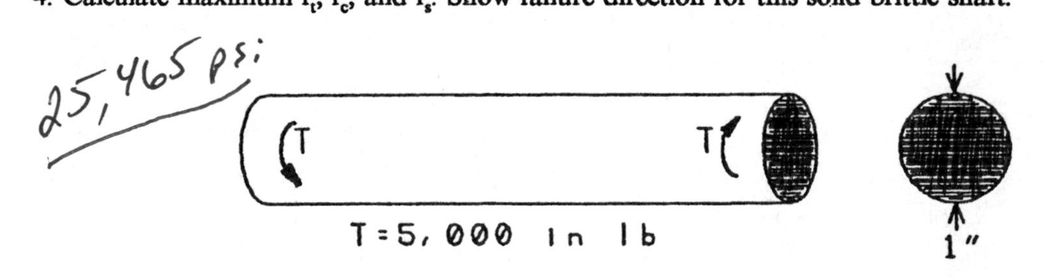

# CHAPTER SEVEN

# RIVETED JOINTS AND PRESSURE VESSELS

## RIVETED LAP JOINTS

The most common type of riveted joint used in aircraft construction is called the *lap joint*. The single shear lap joint is shown in Figure 7.1(a) and the double shear lap joint is shown in Figure 7.1(b).

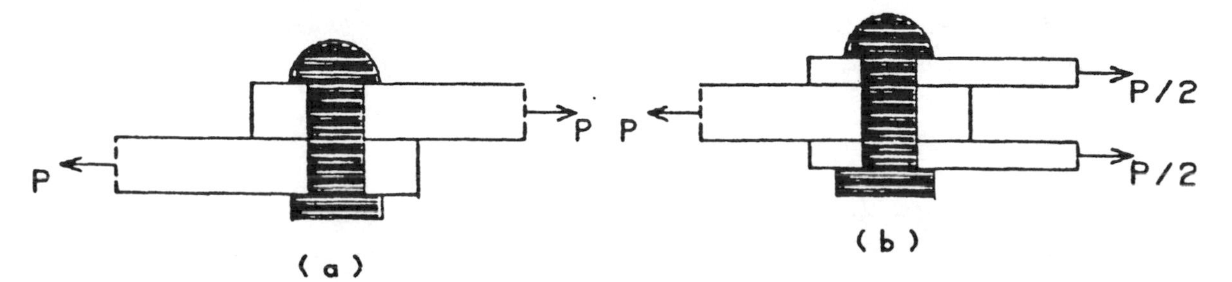


*Figure 7.1 Lap joints: (a) single shear; (b) double shear.*

### Failure Modes

There are four ways that a riveted joint can fail:
(a) rivet shear, (b) tearing (tension) failure of the plate, (c) bearing failure of the plate,
(d) shear tear out of the plate. These are shown schematically in Figure 7.2.

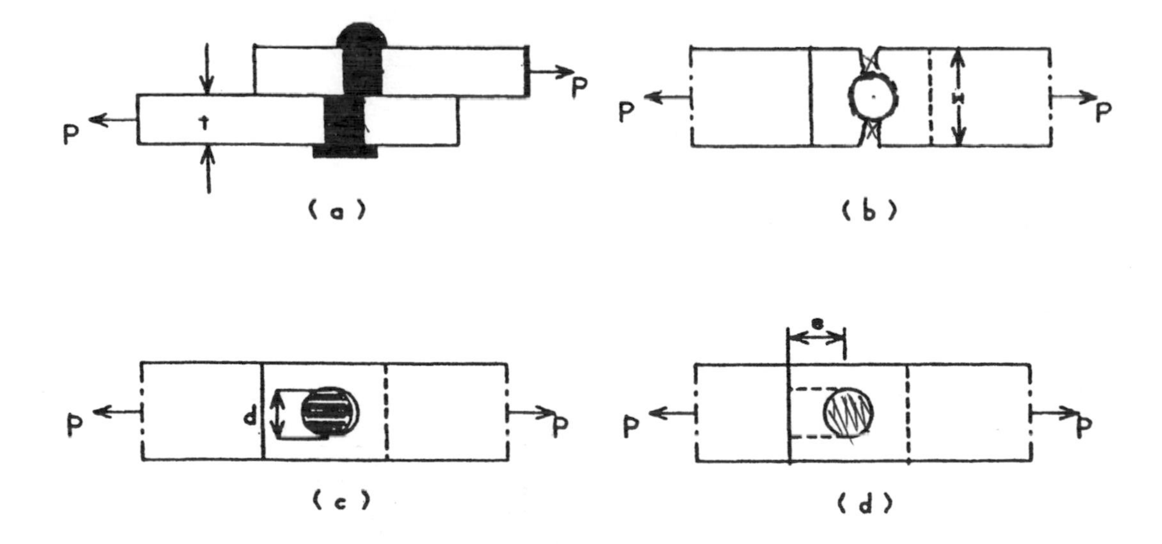


*Figure 7.2 Failure modes: (a) rivet shear; (b) tearing; (c) bearing; (d) shear tear out.*

## Rivet Shear Analysis

If a rivet that is in single shear fails, the broken area, A, will be the cross sectional area of the rivet, $\pi r^2$. The shear stress is calculated by:

$$f_s = \frac{P}{\pi r^2} \tag{7.1}$$

A rivet that is in double shear has two shear areas that resist the shear forces. If shear failure occurs, the broken area is twice as large as in single shear, thus the shear stress is half that of the single shear fastener.

## Tearing (tension) Failure Analysis

Analysis of tearing failures of a row of rivets is simplified by considering a single rivet section as shown in Figure 7.3. The units are: w = width of the section; d = diameter of the rivet hole; t = thickness of the plate. The broken area is t(w-d). The tension stress is:

$$f_t = \frac{P}{t(w-d)} \tag{7.2}$$

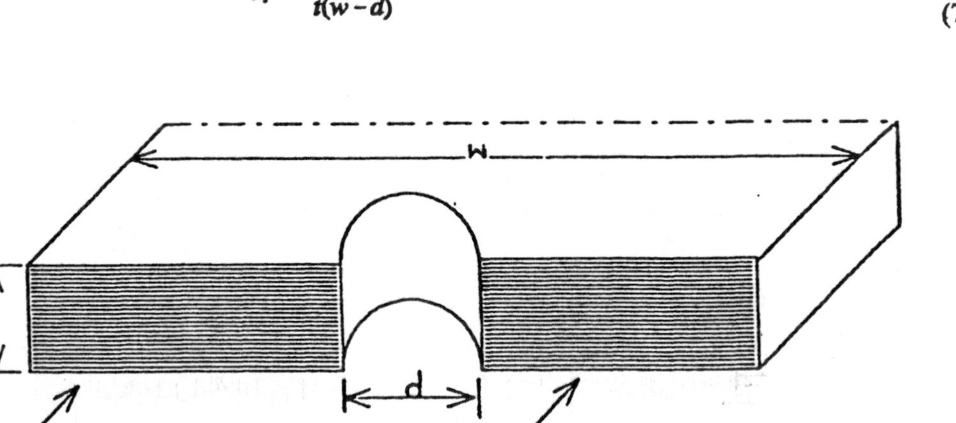


***Figure 7.3 Tearing (tension) failure.***

If analysis shows that a rivet will shear, the logical solution would be to use a larger diameter rivet. But, this will increase the dimension, d, and may reduce the material in the plate to the point that tearing failure will occur.

Another solution would be to use a rivet made of a material with a higher ultimate shear stress. This new material will be harder than the plate material and could cause the third type of failure, bearing failure.

## Bearing Failure Analysis

Unlike the other riveted joint failures, bearing failure occurs without actual separation of the joint. This type of failure results in the elongation of the rivet hole. If the rivet material is harder than the plate material a tension load will cause the plate material toward the edge of the plate to be crushed. The hole will then be elliptical in shape (see Figure 7.4) and the rivet will be loose in the hole.The joint will then be unsatisfactory and will have failed. The compressive stress caused by the harder rivet acting on the softer plate is called the bearing stress, $f_{br}$. The allowable bearing stresses of metals can be found in engineering handbooks. The projected area, A, upon which the load, P, acts upon = dt.

The bearing stress formula is:

$$f_{br} = \frac{P}{dt} \tag{7.3}$$

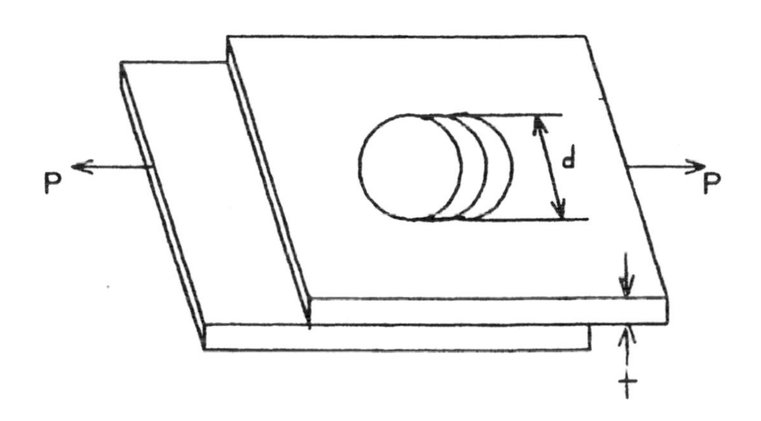


*Figure 7.4 Bearing failure.*

## Shear Tear Out Analysis

Shear tear out (edge tear out) occurs when the rivet holes are drilled too close to the edge of the plate. A good rule of thumb is that the distance from the center of the hole to the edge of the plate, s, should be at least 1.75d. Figure 7.5 shows that two shear areas exist if shear tear out occurs. Each area is equal to s times the plate thickness, t. The shear stress, $f_s$ is then:

$$f_s = \frac{P}{2st} \tag{7.4}$$

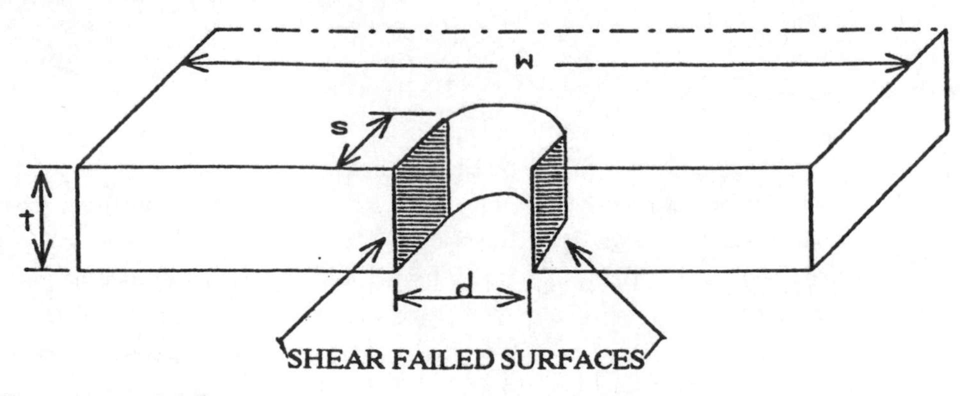


*Figure 7.5 Shear tear out failure.*

## PRESSURE VESSELS

Hydraulic and pneumatic accumulators and pressure lines (tubes) are the usual pressure vessels found in aircraft structures. They are considered to be thin walled shells. The stresses in the wall material caused by internal fluid pressure are considered uniformly distributed. A thin walled vessel is one in which the inside diameter is at least ten times the wall thickness.

### Spherical Pressure Vessels

Consider a spherical metal pressure vessel with internal radius, r, and with uniform thickness, t. It is filled with a compressed fluid with a pressure of, p, pounds per square inch. The schematic cross section of the sphere is shown in Figure 7.6(a). Consider that the sphere is cut through the middle shown in Figure 7.6(b). The total force acting on the hemisphere, P, is equal to the internal pressure, p, multiplied by the cross sectional area, $\pi r^2$ of the cut section. The area of the metal that resists this force, A, is the circumference of the sphere, $2\pi r$, multiplied by the wall thickness, t. The tensile stress of the metal, $f_t$ is:

$$f_t = \frac{P}{A} = \frac{p\pi r^2}{2\pi rt} = \frac{pr}{2t} \tag{7.5}$$

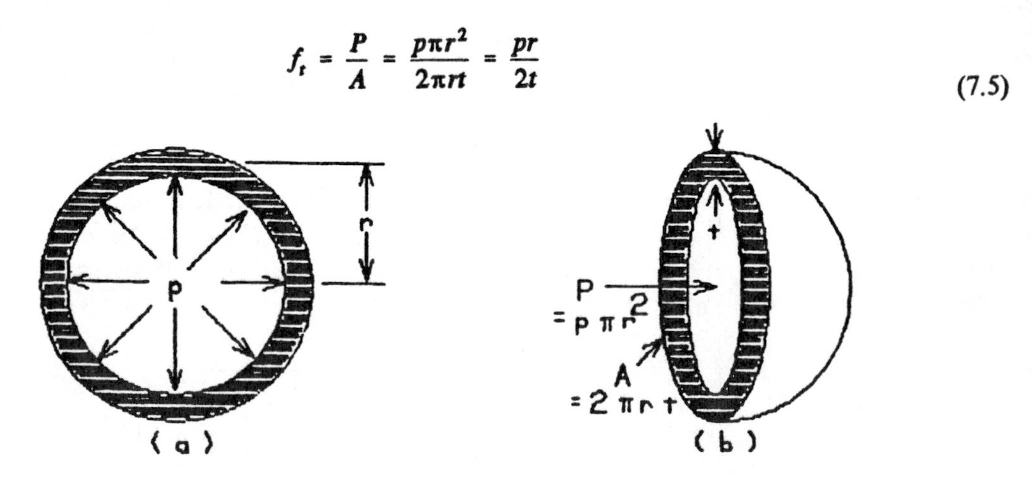


*Figure 7.6 Stress in thin walled sphere.*

## Pressure Lines (Tubes)

Hydraulic and pneumatic pressure lines (tubes) are analyzed as thin walled cylinders. The pressure line shown in Figure 7.7 can fail due to overstress in the longitudinal direction, $f_x$, which would result in a crack along the circumference of the line shown in Figure 7.7(a), or it can fail due to overstress in the direction of the circumference, $f_y$, which would result in a crack in the longitudinal direction shown in Figure 7.7(b).

We will now analyze both stresses to find the actual crack direction.

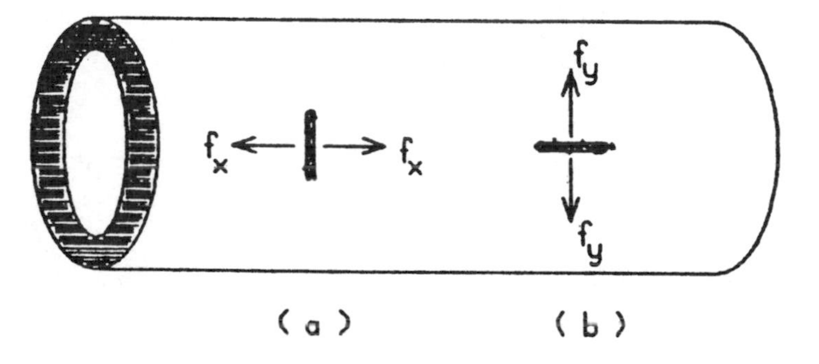


*Figure 7.7 Possible failures in a pressure line.*

Stress in the longitudinal direction is analyzed by cutting across the tube and calculating the applied force, P, on the cut section and the resisting metallic area, A, shown in Figure 7.8. This reveals that the force is the pressure, p, multiplied by the area of the tube, $\pi r^2$. The metallic area is the circumference, $2\pi r$, multiplied by the wall thickness, t. The tension stress of the metal, $f_x$, is:

$$f_x = \frac{P}{A} = \frac{p\pi r^2}{2\pi rt} = \frac{pr}{2t} \tag{7.5}$$

This is the same equation that was derived for the sphere, so we assign it the same equation number, 7.5.

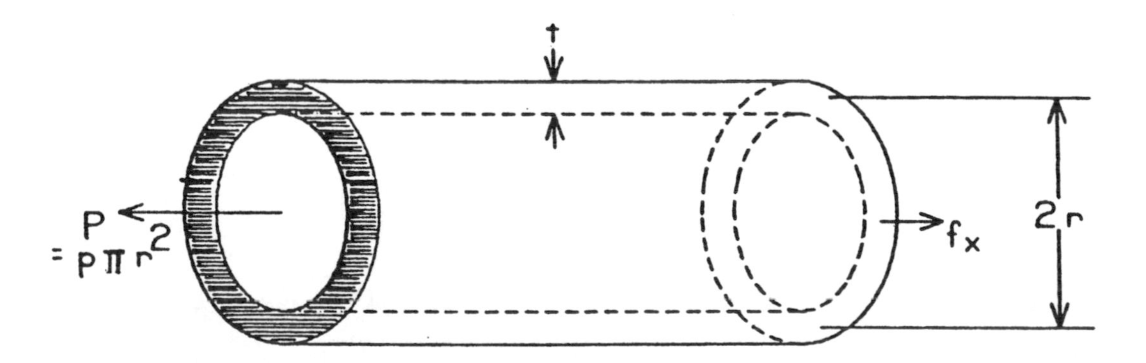


*Figure 7.8 Longitudinal stress in a pressure line.*

Figure 7.9 represents a portion of a pressure line (tube) containing fluid under a pressure of p pounds per square inch. The length of the portion is, L, in., the diameter of the line is, 2r, in., the thickness of the metal line is, t, in.. The total force, P, equals the pressure, p, multiplied by 2rL. The resisting metal area, A, = 2Lt. The tension stress in the direction of the circumference of the line, $f_y$, is calculated by:

$$f_y = \frac{P}{A} = \frac{p2rL}{2Lt} = \frac{pr}{t} \tag{7.6}$$

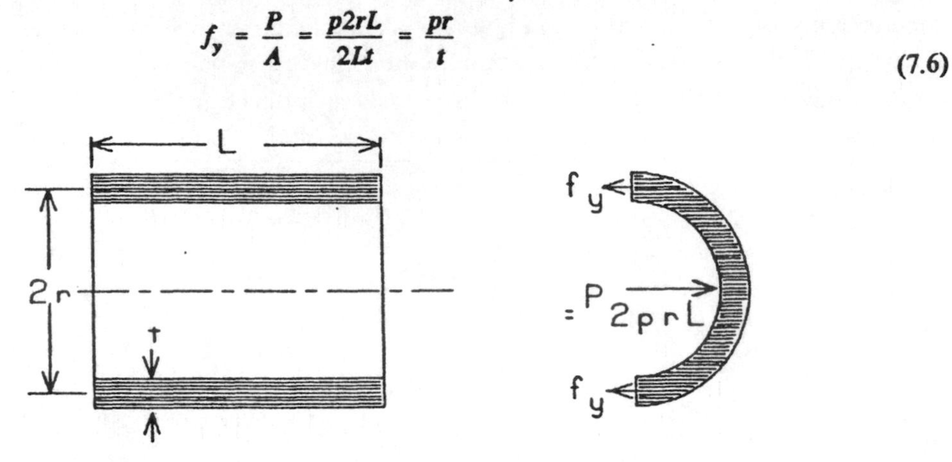


***Figure 7.9 Lateral stress in a pressure line.***

The stress in the direction of the circumference of the pressure line is twice as large as the longitudinal stress. This is called *hoop tension stress.* Thus breaks in pressure lines caused by internal pressure will be in the longitudinal direction shown in Figure 7.10(a).

If breaks occur in the lateral direction they probably will be caused by fatigue cracks or crimping of the line at sharp corner bends shown in Figure 7.10(b).

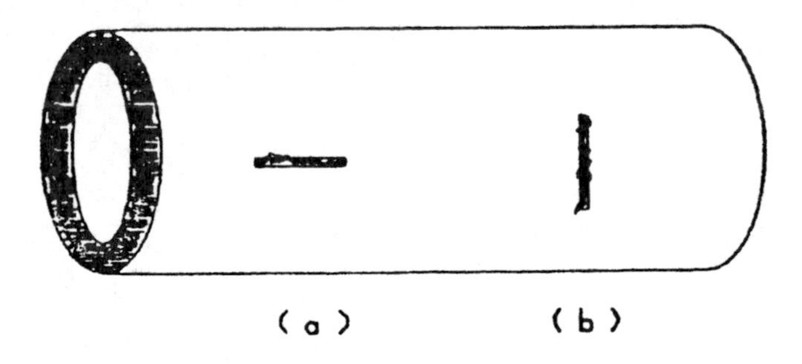

***Figure 7.10 Pressure line breaks.***

Early automobile tires were made with the reinforcing cords in the direction of the circumference of the tire. This is opposite to the direction of the hoop tension stress and blow-outs were common. Later the cords were aligned on the 45° angles, which was a vast improvement. Today's radial tires have cords in the direction of the hoop tension stress and blow-outs seldom occur.

# SYMBOLS, EQUATIONS

## SYMBOLS

$f_{br}$ = bearing (compression) stress, psi.
$f_s$ = shear stress, psi.
$f_t$ = tension stress, psi.
$f_x$ = longitudinal (tension) stress, psi.
$f_y$ = lateral (tension) stress, psi.

## EQUATIONS

### Riveted Lap Joints

7.1 $f_s = \frac{P}{\pi r^2}$ (rivet shear failure)

7.2 $f_t = \frac{P}{t(w-d)}$ (tearing failure)

7.3 $f_{br} = \frac{P}{dt}$ (bearing stress failure)

7.4 $f_s = \frac{P}{2st}$ (shear tear out failure)

### Pressure Vessels

7.5 $f_x = \frac{P}{A} = \frac{p\pi r^2}{2\pi rt} = \frac{pr}{2t}$ (spherical vessels or longitudinal stress in pressure lines)

7.6 $f_y = \frac{P}{A} = \frac{p2rL}{2Lt} = \frac{pr}{t}$ (hoop tension stress in pressure lines)

## REVIEW PROBLEMS

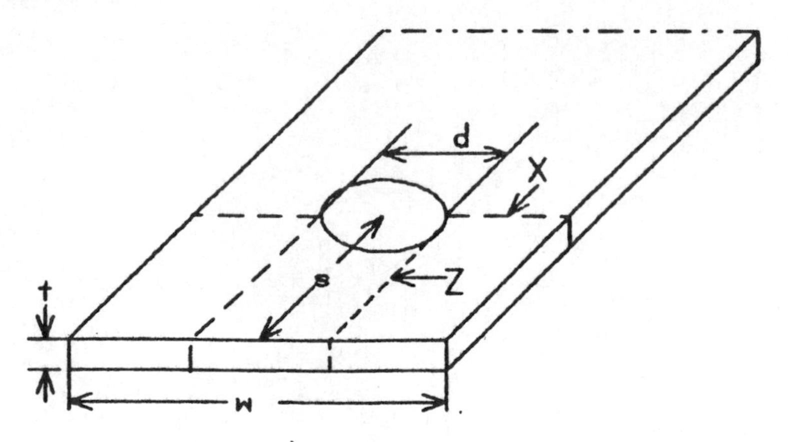


1. If this riveted joint fails on surface X, the failure is:
   (a) a tension failure
   (b) a shear failure
   (c) a bearing failure
   (d) shear tear out

2. If the above joint fails on surfaces Z, the failure is:
   (a) a shear failure
   (b) a tension failure
   (c) caused by dimension "s" being too large
   (d) caused by dimension "t" being too large

3. If the rivet fails in shear the failure is due to:
   (a) the rivet being too close to the edge of the plate
   (b) the rivet is too small in diameter
   (c) the dimension "t" is too small
   (d) the dimension "w" is too small

4. "Hoop tension stress" is:
   (a) twice that of the circumferential stress
   (b) equal in value to the longitudinal stress
   (c) twice that of the longitudinal stress
   (d) none of the above

5. The below sketch is a hydraulic line that has failed due to overpressure. Which letter represents the break? (A)

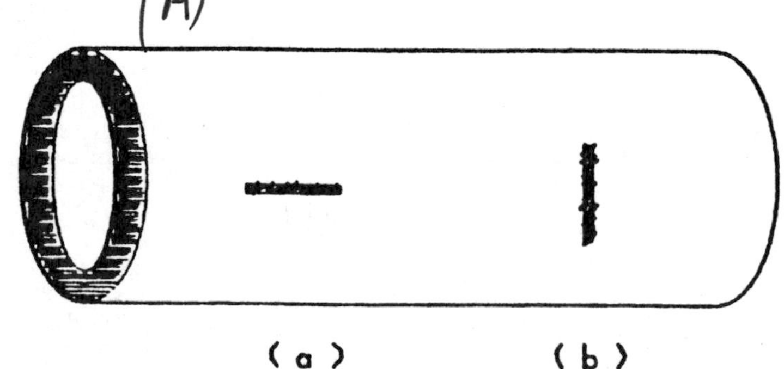

# CHAPTER EIGHT

# STRAIN AND THE STRESS-STRAIN DIAGRAM

*Strain*, $\epsilon$, is the ratio of the deformation of a material caused by an applied load to the material's original size. It is measured by dividing the total deformation by the total dimension of the material in the direction of the deformation. Strain units are inches per inch.

## STRAIN FROM TENSION AND COMPRESSION LOADS

*Axial strain*, $\epsilon_a$, is the change in length, **per unit length**, of a member that is in a tension or compression stressed condition. A distinction must be made between the term "strain" and the term "elongation". *Elongation* refers to the **total** change in length.

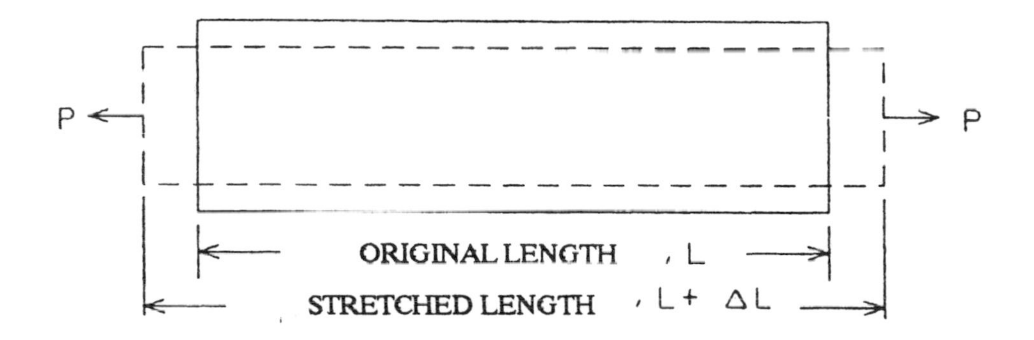


***Figure 8.1 Elongation of member under tension load.***

The strain of the above member is:

$$\epsilon_a = \frac{\Delta L}{L} \tag{8.1}$$

where $\epsilon_a$ = axial tension strain, in/in.
$\Delta L$ = change in length, in.
$L$ = original length, in.

*Lateral strain*, $\epsilon_l$, always occurs with axial strain and always has the opposite sign. The number of molecules in the member shown above does not change when elongation occurs, thus the width of the member must decrease.

This relationship was discovered in 1829 by a French mathematician named Poisson (pronounced pwa-sahn). It is called Poisson's ratio, $\mu$ (mu), and is shown in Figure 8.2.

## Poisson's Ratio

$$\mu = \left|\frac{e_l}{e_a}\right| \approx \frac{1}{3} \tag{8.2}$$

The value of $\mu$ is about the same for most crystalline structural metals.

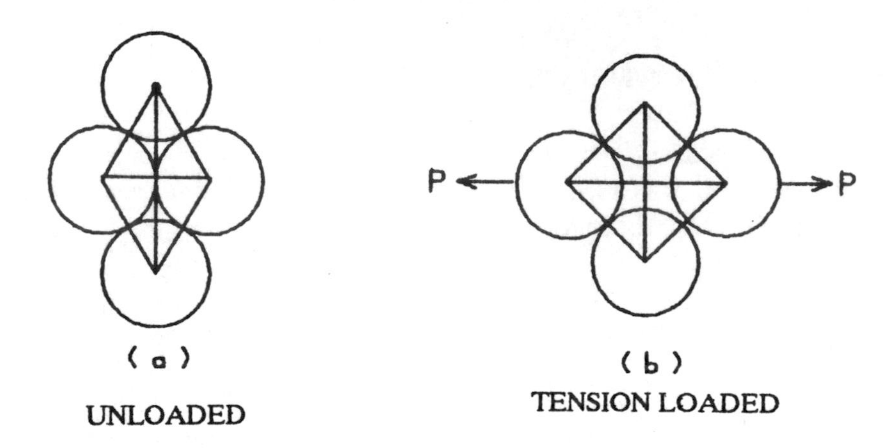

***Figure 8.2 Model for Poisson's ratio.***

If a load, P, is applied to a square steel bar of cross section width, w, it will stretch from a length L to L + ΔL, and the cross section width will decrease from w to w - Δ w, as shown in Figure 8.3.

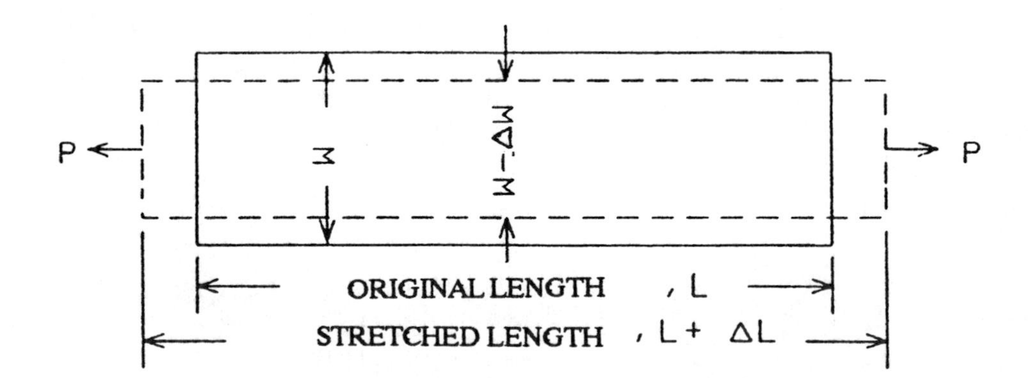

***Figure 8.3 Schematic of axial and lateral strain.***

## MODULUS OF ELASTICITY

If a test specimen of a material is subjected to a tension or compression load and the stress and strain are measured and are proportional to each other, the material is said to be elastic. When the load is removed an elastic material will return to its original dimensions. This property was discovered by an English scientist, Robert Hooke, in about 1658. Some 150 years later another English scientist, Thomas Young, calculated the constant of proportionality, E, called the *Modulus of Elasticity.*

$$E = \frac{f}{\epsilon} \qquad (8.3)$$

where E = modulus of elasticity, psi or in-lb/in$^3$.
f = stress, psi.
$\epsilon$ = strain, in/in.

The modulus of elasticity is also known as *Young's modulus,* and is an important property of a material. It measures the stiffness of the material. Large values of E show that the material is stiff and will have little strain for a given value of stress.

Typical values for some materials used in aircraft are:

| | |
|---|---|
| Steel | E = 30,000,000 psi. |
| Titanium Alloy | E = 17,000,000 psi. |
| Aluminum Alloy | E = 10,000,000 psi. |
| Magnesium Alloy | E = 7,000,000 psi. |

From these figures it shows that, under a certain stress level, aluminum alloy will strain three times as much as steel.

## Shear Strain

*Shear strain,* $\gamma$ ("gamma"), does not measure the ratio of a change in length as axial strain does. Consider the drawing of a square of material under shear loads as shown below.

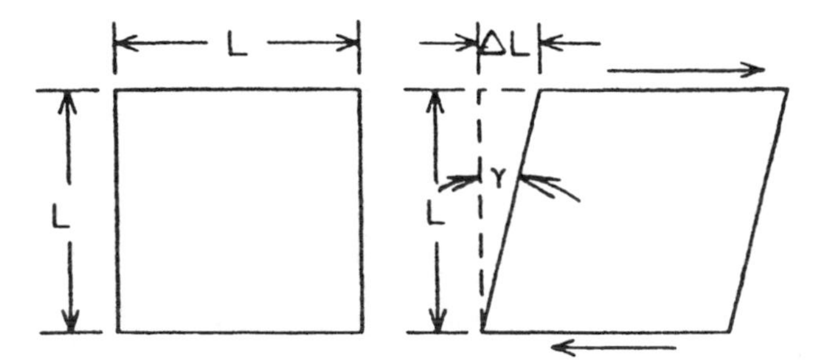


*Figure 8.4 Shear strain (a) unloaded; (b) shear loaded.*

### *Strain and the Stress-Strain Diagram*

Shear loads distort a square into a rhombus as shown in Figure 8.4. Shear strain is the small angle, $\gamma$, shown in the figure. The tangent of $\gamma$ is $\Delta L/L$. Since the angle is small and is measured in radians, the tangent of the angle and the angle are equal. So:

$$\gamma = \frac{\Delta L}{L} \tag{8.4}$$

In shear, instead of modulus of elasticity, we have an equivalent quantity called the *Modulus of Rigidity* or the *Shear Modulus*, G.

$$G = \frac{f_s}{\gamma} \tag{8.5}$$

There is a relationship between the modulus of rigidity, G, and the modulus of elasticity, E as :

$$G = \frac{E}{2(1+\mu)} \tag{8.6}$$

Where $\mu$ = Poisson's ratio = ⅓ . Thus, for crystalline metals, G = 0.375 E

## TENSILE STRESS-STRAIN DIAGRAM

Relationships between stress and strain can be investigated by a tension test. It is convenient to use the original cross sectional area of the test sample in computing the stress of the material. This is called the engineering stress. The actual stress depends upon the instantaneous value of the cross sectional area. This area decreases as tension stress increases (Poisson's ratio).

In the tension test, load, P, applied by a testing machine is converted into stress units by, $f = P/A$. A stress-strain curve is plotted from the results of the test.

## Low Carbon Steel Tensile Stress-Strain Diagram

Although low carbon steel is not used as an aircraft material, the examination of the stress-strain diagram for this material clearly shows significant points on the curve. These cannot be seen for other metals.

### Elastic Behavior

Steel is a highly elastic material. At stress values as high as 75 percent of the ultimate tensile strength, it remains elastic. In this elastic range, if the load is removed the strain disappears completely. Thus a straight line will result on the stress-strain curve (see "0a" on Figure 8.5).

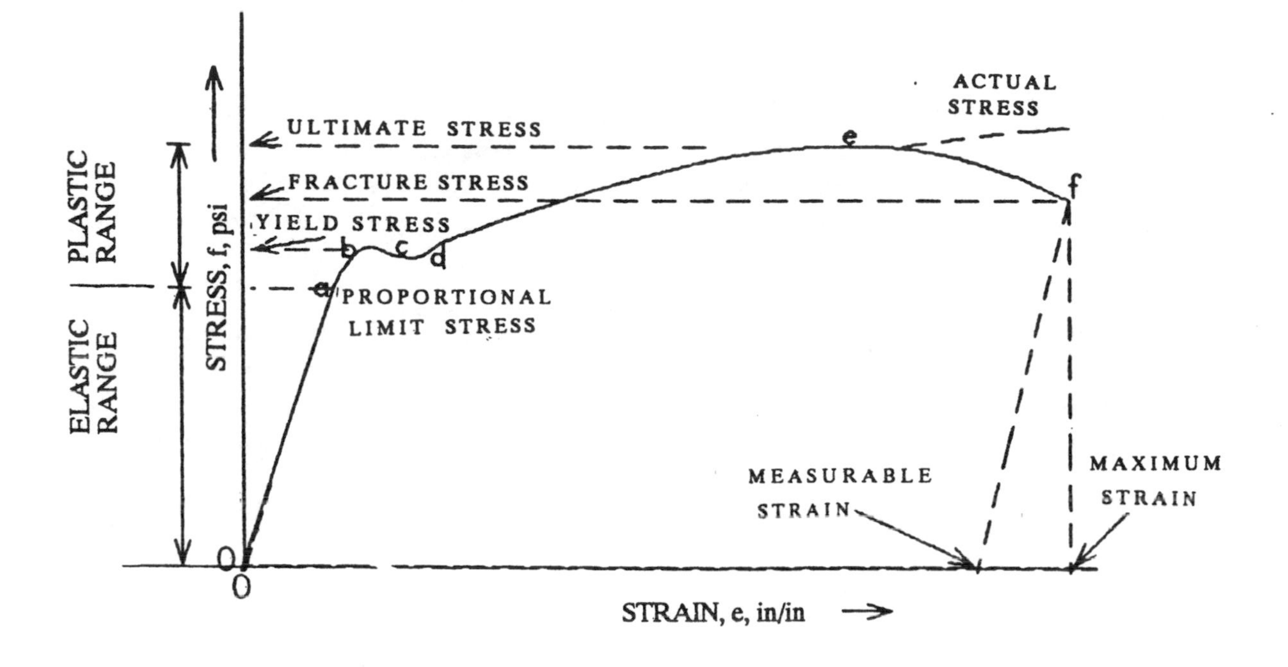


*Figure 8.5 Low carbon steel stress-strain curve.*

## Plastic Behavior

At some stress level, called the *proportional limit stress*, the stress-strain curve changes from a straight line to a curved line. Stress and strain are no longer proportional to each other above this stress level. The metal is no longer completely elastic and, if the load is removed, the metal will not return to the 0-0 point, but will have a permanent strain. This point is shown by the letter "a" on Figure 8.5. All stresses above "a" are in the *plastic range*.

Increasing the load will increase the stress on the metal until point "b" is reached on the stress-strain diagram. At this point a rapid increase in strain occurs and the stress of the material decreases (to point "c"). The metal is said to have reached its *yield stress* at point "b". Metallic slip is occurring as the dislocations move (see Chapter Four). The metal continues to sustain the reduced stress level until all the dislocations that are oriented perpendicular to the applied load have moved to their final positions.

The atomic arrangement is rearranged in a perfect alignment. The metal can now sustain greater stress levels and the curve again resumes its upward direction at point "d".

## *Strain and the Stress-Strain Diagram*

The stress level continues to increase as more load is applied until the ultimate engineering stress is reached at point "e". It must be reemphasized that engineering stress is not the actual stress, because it is calculated by using the original cross sectional area, not the actual instantaneous area. Actual stress will continue to rise until fracture as shown in Figure 8.5.

The load cannot be sustained beyond this point and the engineering stress decreases until fracture occurs at point "f". For ductile metals, such as this low carbon steel, there is a "necking down" or noticeable reduction in cross section before failure occurs. Necking does not occur until the ultimate stress point is reached. The fracture surface of a round low carbon steel test specimen exhibits a brittle type failure at the center and ductile type 45° "shear lips" at the edges. This failure is shown in Figure 8.6(a).

Brittle metals, such as high carbon steel, are comparatively weak in tension. They do not neck down as ductile materials do, nor do they exhibit shear lips. They fail on a 90° plane to the applied tension load as shown in Figure 8.6(b).

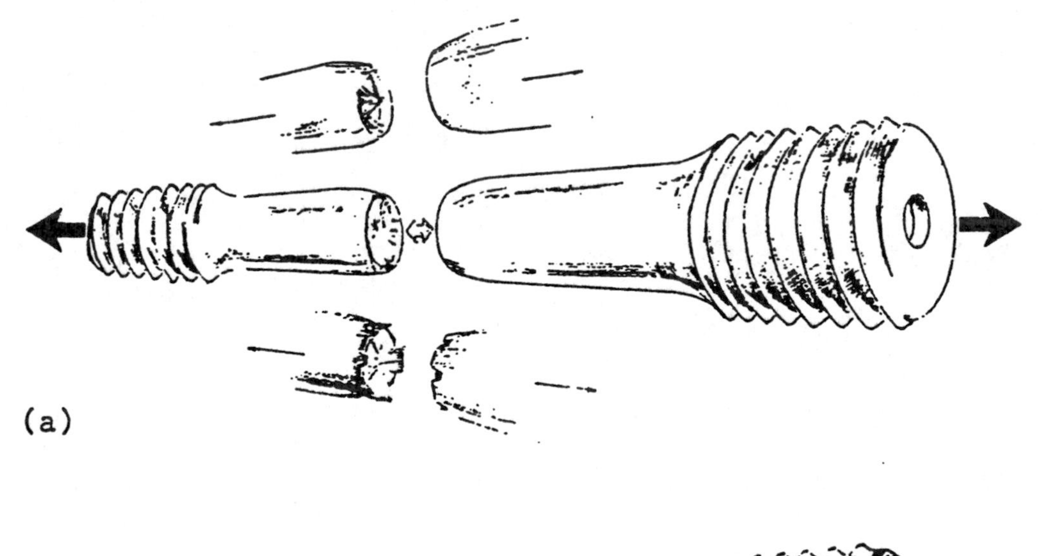


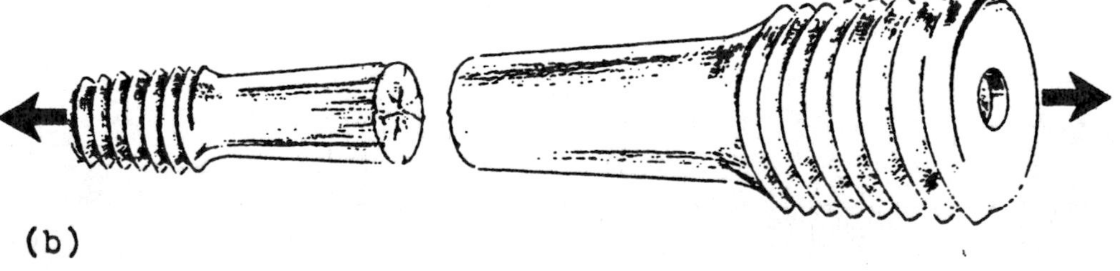

*Figure 8.6 Failed round tension test specimens, (a) low carbon steel; (b) high carbon steel.*

## Maximum Strain

The maximum strain that a metal can sustain before failing in tension is a measure of its *ductility.* High strain values are reached in ductile materials before failure but brittle materials fail at low strain values. Ductility is measured in percent of original length instead of actual strain values. Thus ductility is merely the strain multiplied by 100.

## Modulus of Elasticity

The modulus of elasticity, E, is equal to the stress divided by the strain as shown in Equation 8.3.
For materials with stresses in the elastic range the value of E will be a constant. To find the value, divide the proportional limit stress by the corresponding strain. This is the tangent of the angle of the stress-strain curve in the elastic range. This is called the "slope of the curve." E is shown as an angle in Figure 8.7(a).

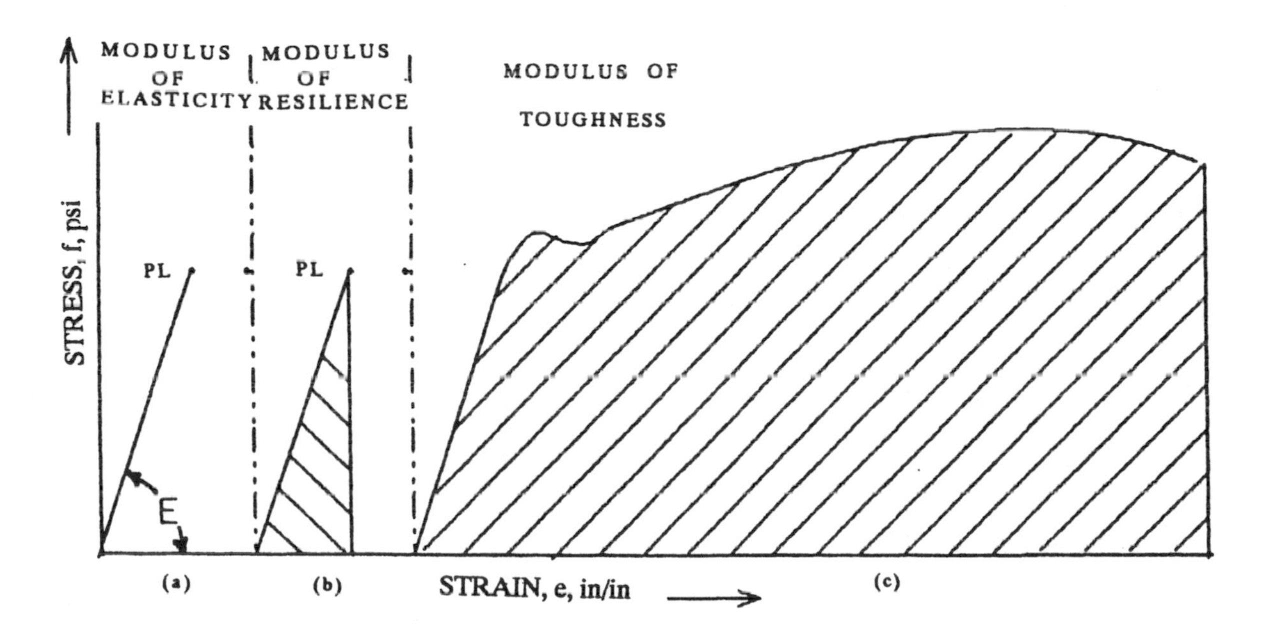


## Modulus of Resilience

The *Modulus of Resilience,* more simply called *resilience,* is a measure of the amount of energy a material can absorb elastically in a unit volume of the material. This energy is not wasted as heat, and can be recovered.
High resilience is desired in watch springs where the energy absorbed in winding the spring can be recovered to run the watch.

The amount of resilience is measured by the area of the triangle under the straight part of the stress-strain curve (Figure 8.7(b)). The units of resilience are those of stress (psi) x strain (in/in). These are psi or in-lb/in$^3$.

## Modulus of Toughness

*Toughness* of a material is the total energy absorbed before failure occurs. It is shown by the entire area under the stress-strain curve (Figure 8.7(c)).

For some applications, e.g., sledge hammers, pile drivers, and crash helmets, toughness is a desired quality.

For simplicity, toughness can be measured by assuming that the area under the curve is a rectangle. The base of the rectangle is equal to the total strain and the height is equal to the average stress.

Units of toughness are psi or in-lb/in$^3$.

## Summary of Stress-Strain Diagram Properties

*Proportional limit stress,* is the highest stress that a material can withstand while remaining elastic.

*Yield stress,* is the stress at which objectionable permanent strain occurs.

*Ultimate stress* (or ultimate strength), is the highest engineering stress that the material can withstand before fracture.

*Fracture stress,* is the engineering stress at the point of fracture.

*Elastic range,* is the stress range from zero to the proportional limit stress.

*Plastic range,* is the stress range from the proportional limit stress to the ultimate stress.

*Maximum strain,* is the total strain at the point of fracture. It is a measure of the ductility of the material.

*Modulus of Elasticity,* is a measure of stiffness of a material. It is calculated by the slope of the straight line (elastic range) portion of the diagram.

*Modulus of Resilience,* is a measure of the ability to store energy elastically. It is calculated by the area under the straight line (elastic range) portion of the diagram.

*Modulus of Toughness,* is a measure of the ability to absorb energy before failure. It is calculated by the entire area under the diagram.

## Work-Hardening(Strain-Hardening)

If a metal specimen is loaded in tension beyond its proportional limit and then is unloaded the unloading curve will not retrace the loading curve. The unloading curve will be parallel to the straight line portion of the loading curve.

This is shown in Figure 8.8 where unloading took place at point "x". When the load (and stress) has returned to zero there will be a permanent strain of the metal ("0y").

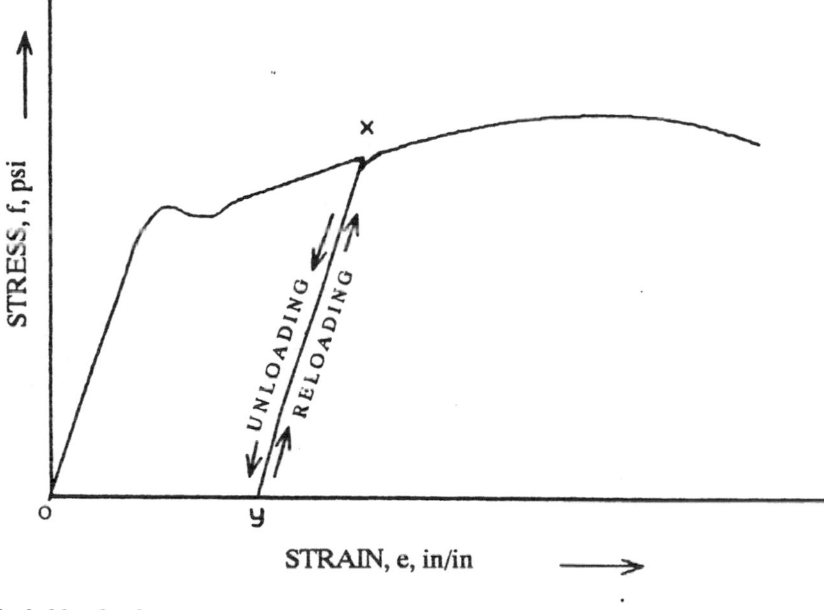


*Figure 8.8 Work Hardening.*

If the specimen is reloaded in the same direction the new stress-strain curve will return to point "x" by the same path that was followed by the unloading curve. The curve will then continue as though no interruption to the original loading took place.

This phenomenon is called *work-hardening, strain-hardening* or *cold-working*. If the stress at point "x" is greater than the yield stress, the result of this process is an increase in the elastic range, proportional limit, and yield stress. There is no change in the ultimate tensile stress. The ductility of the metal is reduced by the amount of permanent strain (also called "set").

Cracking of the metal and residual stresses may result from work-hardening. So this process is not used above about 90 percent of the metal's ultimate stress. Cracking can lead to metal fatigue. Residual stresses can lead to *stress corrosion cracking*.

## Higher Carbon Steel Diagrams

The low carbon steel tensile stress-strain diagram is changed with an increase in carbon content of the steel. The properties of the steel that change with increased carbon content include: increase in yield stress and ultimate strength. There is a corresponding decrease in ductility. The proportional limit stress and yield stress are less clearly defined. Carbon content has very little effect on the modulus of elasticity so the slope of the elastic portion of the stress-strain diagram is unchanged.

Stress-strain diagrams for 1020 (0.20 percent carbon), 1040 (0.40 percent carbon), and 1090 (0.90 percent carbon) steels are shown in Figure 8.9.

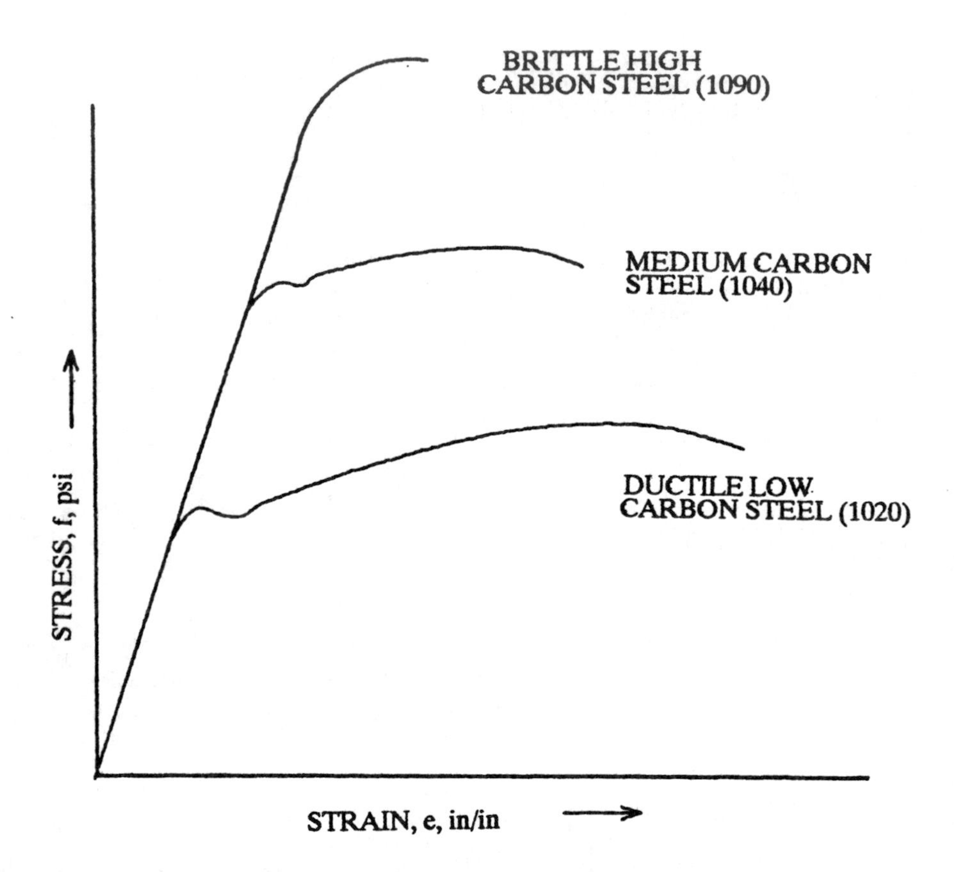


***Figure 8.9 Tensile stress-strain diagrams for varying carbon content steels.***

High carbon steels have a greater resilience than low carbon steels. Resilience is shown by the area under the straight line portion of the curve.
This is shown in Figure 8.10.

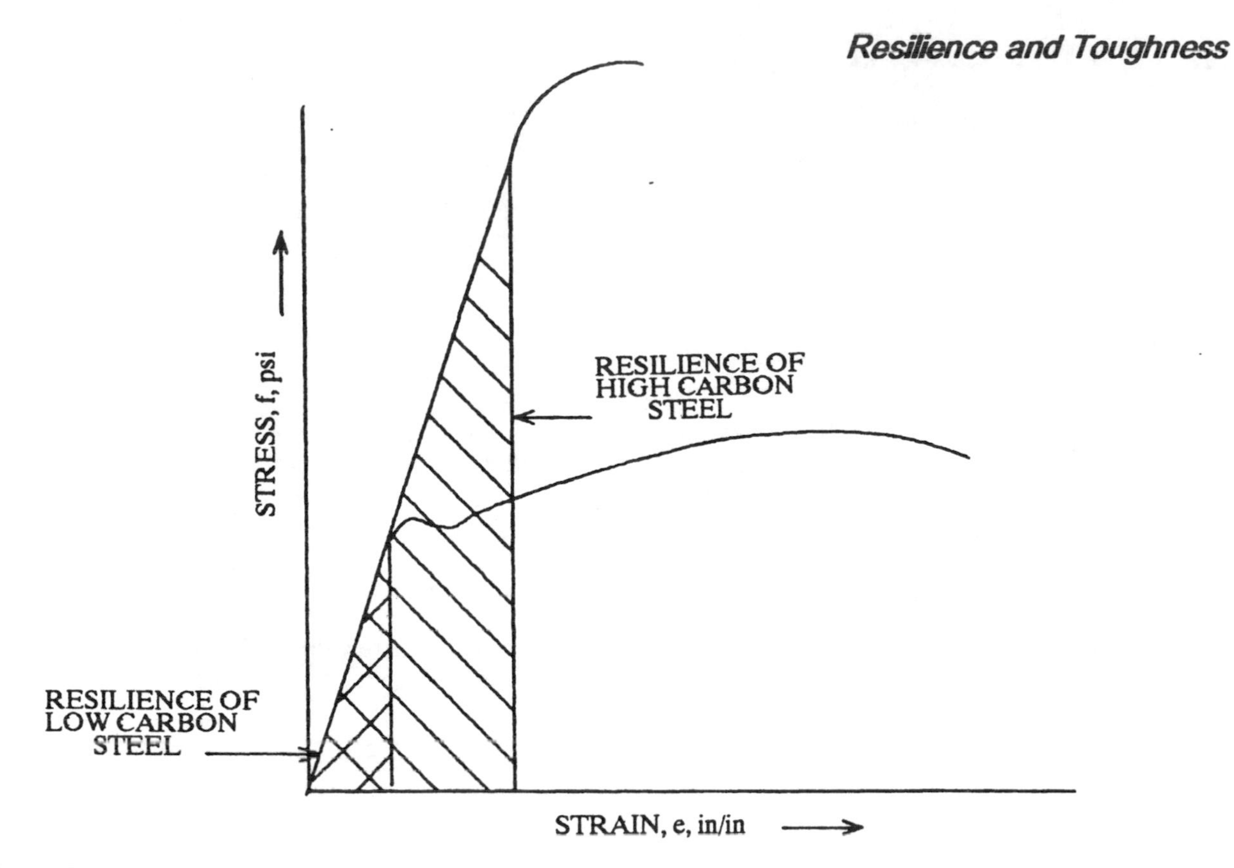


***Figure 8.10 Effect of carbon content in steel on resilience.***

Adding carbon to steel increases the ultimate stress of the steel but reduces the ductility. Toughness is shown by the area under the curve. As shown in Figure 8.11, the area, and thus the toughness, is less in the high carbon steel.

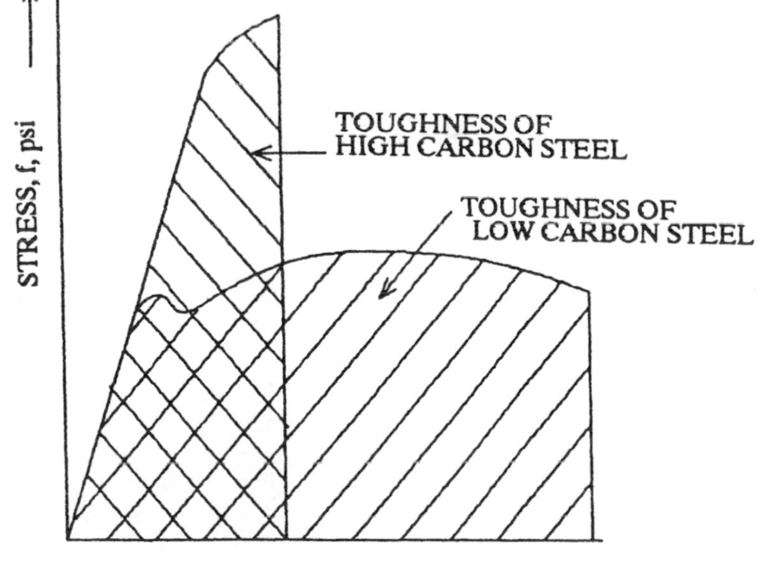


***Figure 8.11 Effect of carbon content in steel on toughness.***

Most of the physical characteristics of steel (and other metals) can be altered by alloying them with other elements. Heat treatment also will change most of the characteristics. We will discuss this later. The modulus of elasticity is one characteristic that neither of these methods will appreciably change.

## Stress-Strain Diagram for Other Materials

Most metals used in aircraft construction, such as aluminum and magnesium, do not exhibit a definite proportional limit stress or yield stress. These values are found by an arbitrary process called the "offset" method. The process consists of: measuring a strain of 0.0001 in./in. on the strain scale, drawing a line parallel to the modulus of elasticity, and marking the intersection of this line and the stress-strain diagram. The stress value of this point is the proportional limit stress.

A similar process is used to find the yield stress. Measuring a strain value of 0.002 in./in. is used in finding the yield stress.

Figure 8.12 illustrates this.

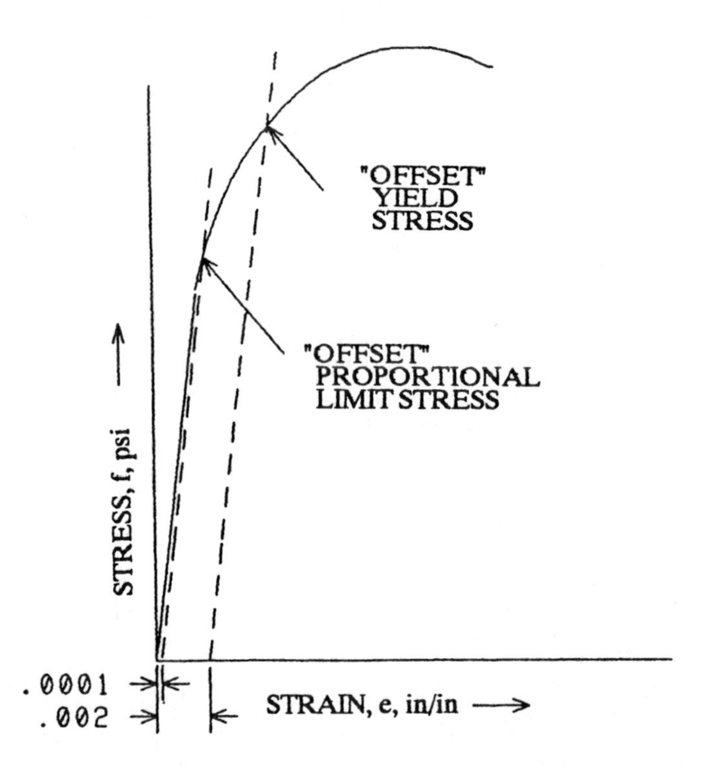

*Figure 8.12 "Offset" method of finding proportional limit and yield stress.*

## ENGLISH SYMBOLS

E = modulus of elasticity, psi.
$\epsilon$ = strain, in./in.
$\epsilon_a$ = axial strain
$\epsilon_l$ = lateral strain
f = stress, psi
G = modulus of rigidity (shear), psi.

## GREEK SYMBOLS

$\gamma$ (gamma) = shear strain, in./in.
$\Delta$ (DELTA) = "change in"
$\mu$ (mu) = Poisson's ratio, dimensionless

## EQUATIONS

8.1 $\epsilon_a = \frac{\Delta L}{L}$

8.2 $\mu = \left|\frac{\epsilon_l}{\epsilon_a}\right| \approx \frac{1}{3}$

8.3 $E = \frac{f}{\epsilon}$

8.4 $\gamma = \frac{\Delta L}{L}$

8.5 $G = \frac{f_s}{\gamma}$

8.6 $G = \frac{E}{2(1+\mu)}$

## REVIEW PROBLEMS

1. What are the units for stress? for strain? psi - in/in

2. According to Poisson's ratio, if the tension strain of a material is 0.3 in./in., what is its lateral strain? .1 in/in

3. A 10 inch long metal bar is placed in tension and its strain is 0.20 in./in. Find the new length. 12 inch

4. The modulus of elasticity of a material measures its: (a)strength; (b)ductility; (c)stiffness; (d)toughness.

5. If a metal is stressed to 15,000 psi and is found to have strained 0.0005 in./in., calculate E.
30, 000, 000 psi, steel

6. Calculate the longitudinal elongation of this steel bar.

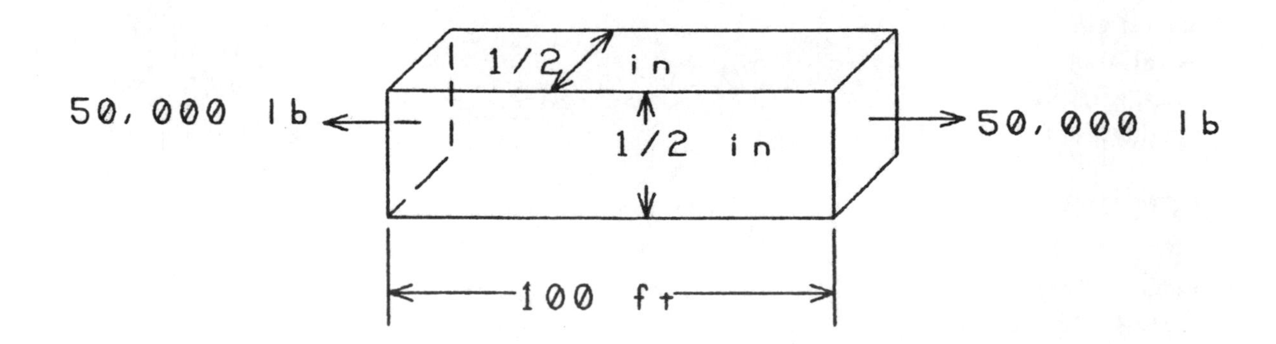


On the figure below:

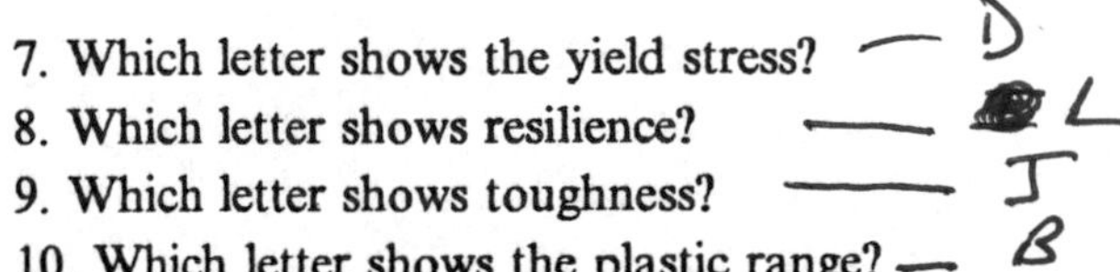

7. Which letter shows the yield stress?
8. Which letter shows resilience?
9. Which letter shows toughness?
10. Which letter shows the plastic range?

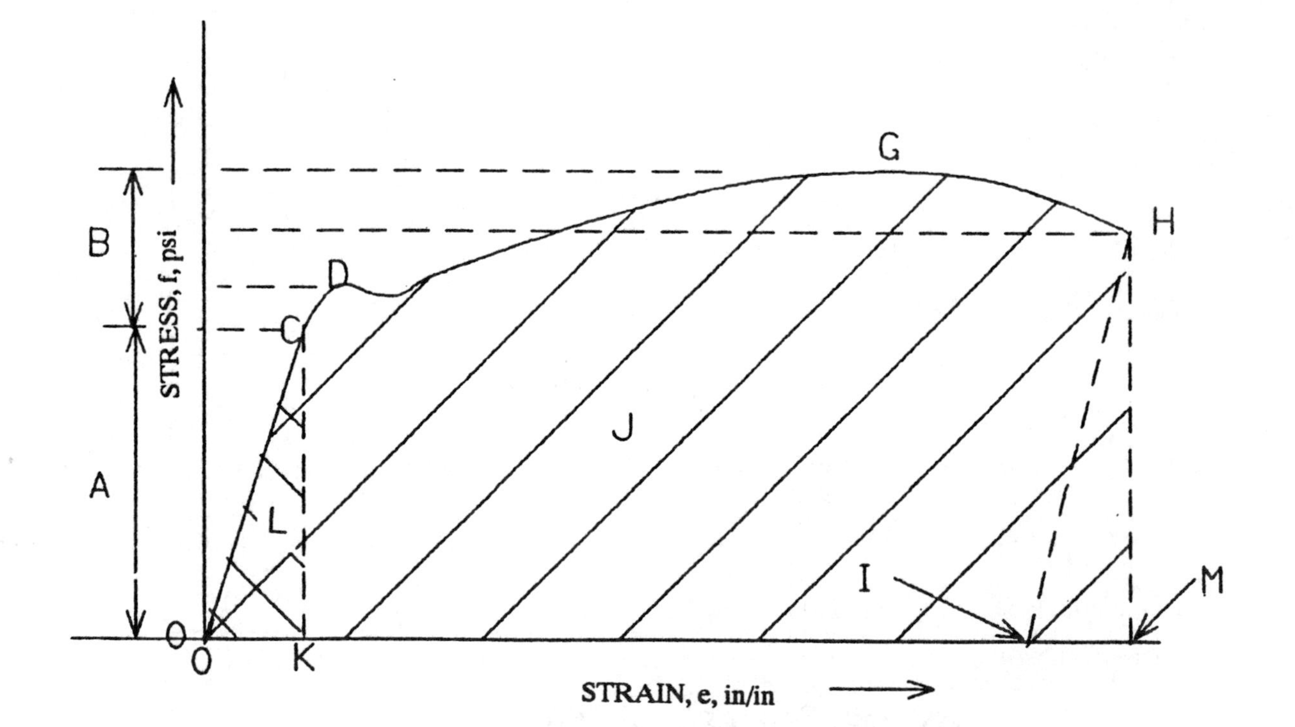

# CHAPTER NINE

# COMPRESSIVE FAILURES

In Chapter Five we calculated compressive stress by formula 5.1, $f_c = P/A$. We also mentioned that, "most compression failures in aircraft structures do not occur because of compression over stress." We will now discuss the types of failure resulting from compression overloads.

The classic case of exceeding the ultimate compressive stress is demonstrated by examining a penny which was placed on a railroad track and run over by a train. If excessive deformation prevents a part from doing its job and it has failed.

Aircraft parts seldom fail from this type of compression failure.

A second type of compression failure occurs in brittle materials. These materials are very strong in compression, but somewhat weak in shear. The type of failure that results from compression loads is from the component shear stress. It occurs on the 45° plane. Brittle materials are seldom used in aircraft, thus this type of failure is not common.

Compression overloading in aircraft structures nearly always results in buckling failure. These failures are discussed as column failures.

## COLUMNS IN COMPRESSION

### **Long Columns** (Illustrations in the Appendix)

If a long straight column is subjected to a gradually increasing compressive load, initially it will remain straight. But at some point the column will start to buckle and it will be unable to support the load. The load at this point is called the *buckling critical load,* $P_{cr}$.

If the load is removed quickly when $P_{cr}$ is reached and the column snaps back to its straight shape. The column failed while still in the elastic range.

If the load is sustained at $P_{cr}$, the column will continue to bend and will not return to its original straight shape after the load is removed.

Failure occurs when $P_{cr}$ is reached. The column is still elastic at this point.

Columns that fail in this manner are called *long columns.*

## **Short Columns** (Illustrations in the appendix)

Another column failure results when the elastic limit of the column is exceeded. There is no apparent bending of this type of column. When $P_{cr}$ is reached there is a sudden permanent buckling of the column. The material is above its elastic limit and in the plastic range. The $P_{cr}$ for this type of failure is much higher than that for the long column with the same cross section dimensions, material, and end restraints. Columns that fail in this manner are called *short columns.*

Figure 9.1 compares short and long columns.

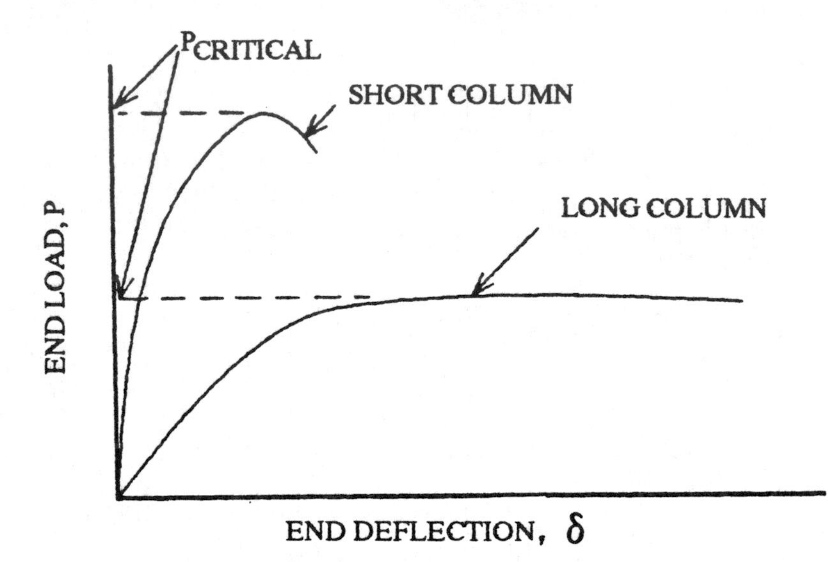


***Figure 9.1 Short and long column critical loads.***

## **Factors Involved in Column Failures**

In the discussion of bending loads in Chapter Six we saw the importance of moment of inertia, I, to bending stresses. A similar effect is found in column buckling. High values of I reduce the buckling stress and help prevent failure. In bending, the moment of inertia about the neutral axis is calculated.

In columns buckling can take place about either of the two principal axes, so the value of I about both axes must be calculated and the lesser value used. Another important parameter is the cross sectional area, A, of the column. The moment of inertia and the cross sectional area are combined into a term named *radius of gyration*, ρ (rho).

$$\rho = \sqrt{\frac{I}{A}} \tag{9.1}$$

Another important factor in column buckling is the manner in which the ends of the column are restrained. This is called the *end-fixity.* End-fixity coefficients, K, range from 1, for a column that is hinged at both ends so that it can rotate freely, to 4, for a column that is rigidly fixed against rotation at both ends.

The *effective length,* L', is:

$$L' = \frac{L}{\sqrt{K}} \tag{9.2}$$

Where L is the actual length of the column. Columns of any classified by their *effective slenderness ratios,* $SR_e$.

$$SR_e = \frac{L'}{\rho} \tag{9.3}$$

For a slenderness ratio greater than a certain critical value, the column is a long column.

A column with a slenderness ratio less than the critical value is a short column. The critical value of the slenderness ratio occurs where the compressive stress in the column is equal to the proportional limit stress of the material.

## Column Stress Formulas

The *Euler formula,* also called the *long column formula,* is valid for columns, which fail in the elastic range.

$$f_c = \frac{\pi^2 E}{(L'/\rho)^2} \tag{9.4}$$

Where: $f_c$ = allowable column compressive stress, psi.

E = modulus of elasticity, psi.

$L'/\rho$ = effective slenderness ratio, dimensionless

The short column fails in the plastic range. The modulus of elasticity is not a constant beyond the elastic range so a new term, *tangent modulus,* $E_t$, is introduced. The tangent modulus is the slope of a tangent to the compressive stress-strain diagram at the given stress. With this modification, the Euler formula becomes:

$$f_c = \frac{\pi^2 E_t}{(L'/\rho)^2} \tag{9.5}$$

The calculated values of stress are based upon the assumption that the column is straight and is loaded at its *centroid* (center of area). If these conditions are incorrect, lower stress values will result.

Allowable column stresses for various effective slenderness ratios are shown in Figure 9.2.

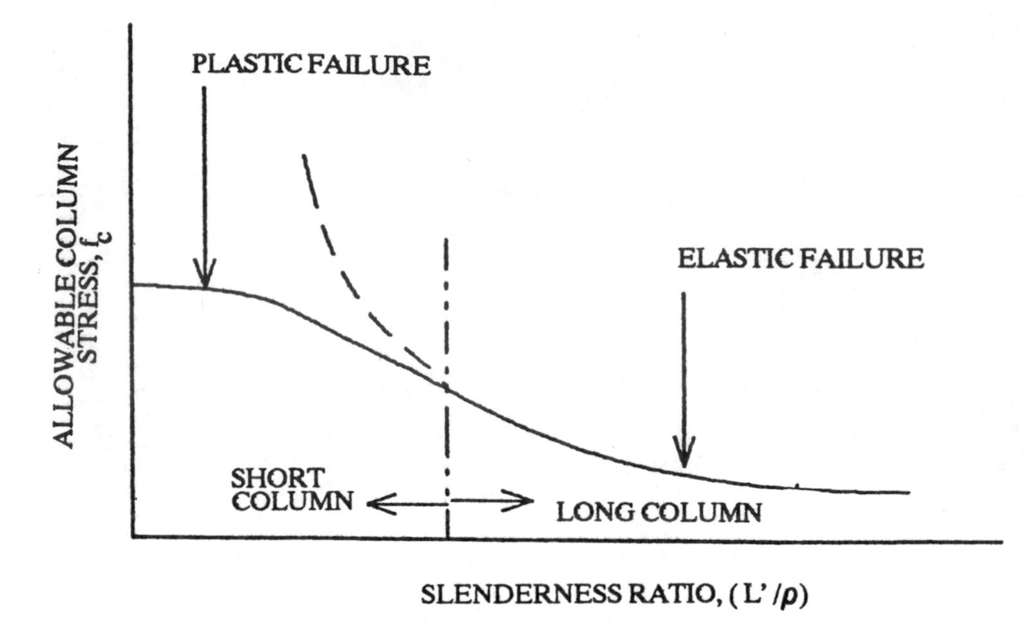


***Figure 9.2 Allowable column compressive stresses.***

## Effect of Column Cross Section

To emphasize the effect of the cross sectional shape of a column in resisting compressive loads, consider a 20 inch long column of rectangular cross section as shown in Figure 9.3(a). The metal is aluminum alloy with a modulus of elasticity of 10,000,000 psi. The dimensions of the cross section are 2.125 x 0.064 in..

The column is pin ended so K = 1, thus L' = L = 20 in. Equation 9.4 can be rewritten with these constants as: $P_{cr}$ = 246,740 I.

The least (smallest) moment of inertia of the column shown in Figure 9.3(a), I = 0.0000464 $in^4$. The critical load, $P_{cr}$ = 11 lb.

Reshaping the same material into an equal leg angle (Figure 9.3(b)), the least moment of inertia, I = 0.002845 $in^4$, and $P_{cr}$ = 702 lb.

Finally, reshaping the material into a Z section with 1/2 inch legs (Figure 9.3(c)), the least moment of inertia, I = 0.005361 $in^4$, and $P_{cr}$ = 1323 lb.

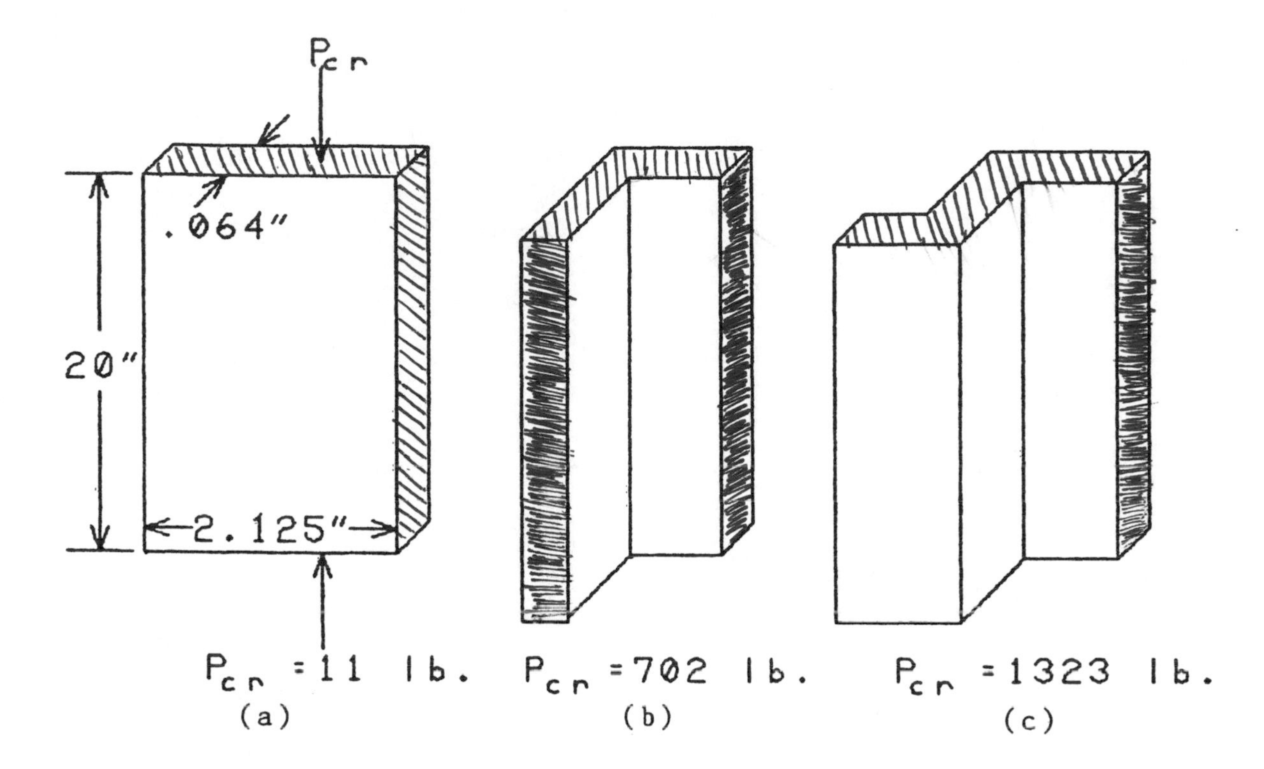


***Figure 9.3 Relative critical loads of equal area columns.***

## PANEL BUCKLING (More illustrations in the Appendix)

Figure 9.4 shows two square panels loaded by a distributed force, P, in compression.

Panel (a) is not restrained at the edges and so acts like a wide column. Formulas 9.4 and 9.5 apply here.

Panel (b) is restrained at the edges. The edge fixity coefficient, K, for restrained edges is found from the table below.

These coefficients are not to be confused with the column end fixity coefficients discussed earlier.

| **Edge Restraint** | **K** |
|---|---|
| All four edges fixed | 6.3 |
| All four edges simply supported | 3.6 |

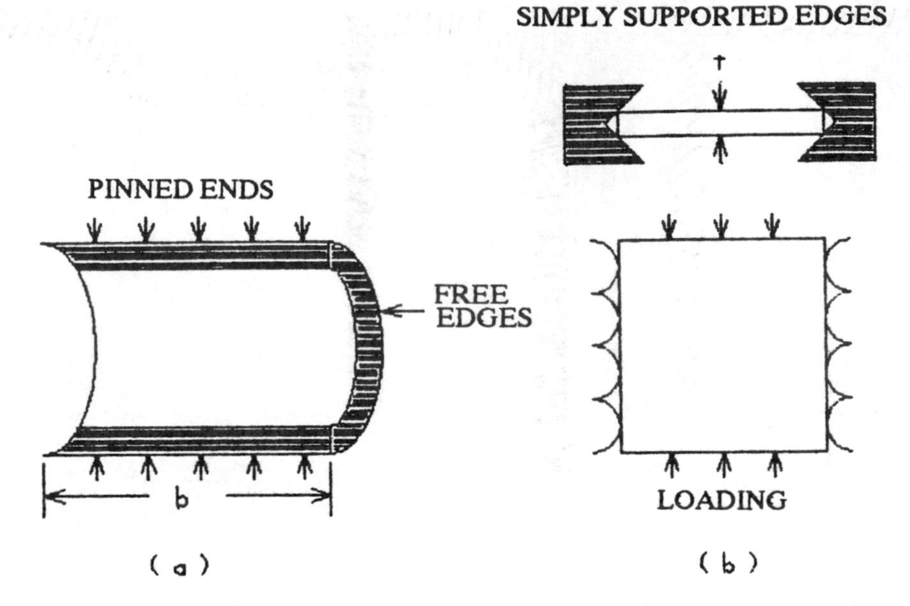


***Figure 9.4 Panel buckling: (a) unloaded edges free; (b) unloaded edges simply supported.***

The critical stress for a panel loaded in compression depends upon the modulus of elasticity of the material, edge restraint, and dimensions of the panel.

The equation is:

$$f_{cr} = KE\left(\frac{t}{b}\right)^2 \qquad (9.6)$$

Where $f_{cr}$ = panel buckling stress, psi.

K = panel edge fixity coefficient

E = modulus of elasticity, psi.

t = panel thickness, in.

b = length of loaded edge, in.

## COMPRESSION FAILURES IN ROUND TUBES (More Illustrations in Appendix)

If a short thin-walled tube (such as an aluminum can) is loaded in compression, the tube will fail by buckling of the tube walls rather than by bending of the tube. The buckling failure will be diamond shaped as shown in Figure 9.5.

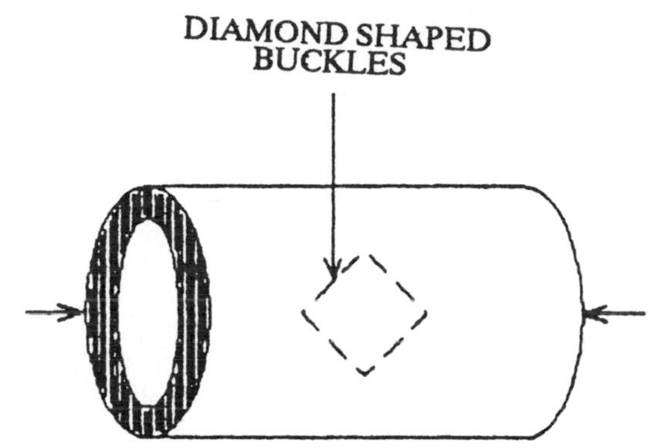


*Figure 9.5 Short tube buckling in compression.*

The critical compression stress, $f_{cr}$, is found by:

$$f_{cr} = K\sqrt{EE_t}\left(\frac{t}{D}\right) \tag{9.7}$$

Where K = constant (0.4, conservative value)
E = modulus of elasticity, psi.
$E_t$ = Tangent modulus of elasticity
t = wall thickness, in.
D = diameter of tube, in.

## ENGLISH SYMBOLS

K = end-fixity coefficient, dimensionless
$E_t$ = tangent modulus of elasticity, psi.
L = actual length of column, in.
L' = effective length of column
$P_{cr}$ = critical load, lb.

## GREEK SYMBOL

ρ (rho) = radius of gyration, in.

## EQUATIONS

9.1 $\rho = \sqrt{\frac{I}{A}}$

9.2 $L' = \frac{L}{\sqrt{K}}$

9.3 $SR_e = \frac{L'}{\rho}$

9.4 $f_c = \frac{\pi^2 E}{(L'/\rho)^2}$ (long column)

9.5 $f_c = \frac{\pi^2 E_t}{(L'/\rho)^2}$ (short column)

9.6 $f_{cr} = KE(\frac{t}{b})^2$ (panel)

9.7 $f_{cr} = K\sqrt{EE_t}\left(\frac{t}{D}\right)$ (tube)

## REVIEW PROBLEMS

1. (True) False. Long columns fail while they are still in their elastic range.

2. True-(False.) Effective slenderness ratio is the factor which determines whether a column is a short column or a long column.

3. True-(False) Long columns are stronger than short columns of the same material and cross section.

4. What do we mean by Tangent Modulus? When do we use Tangent Modulus?
   Slope of tangent to f-e diagram; To find $f_c$ for short column

5. How much stronger in buckling is a panel with all edges riveted than the same panel with simply supported edges? 1.75 times

6. A cylindrical metal fuel tank is found to have diamond shaped buckles. What type of loading caused them.
   Compression

7. A 1 inch x 2 inch rectangular aluminum column is 40 inches long. The column is centrally loaded in compression from its ends. The ends are rigidly restrained from rotating. Calculate:
   (a) radius of gyration,(b) effective length,(c) effective slenderness ratio.(d) long column $f_{cr}$.
   .288675 20 in 69.281465 20,562 psi

   E (aluminum) = 10,000,000 psi., I (rectangle) = $(b)(h)^3$ 12

# CHAPTER TEN

## AIRCRAFT METALS AND THEIR HEAT TREATMENT

In this chapter we will discuss some more common metals used in aircraft construction. Composite materials are discussed in Chapter Eleven. A discussion of how metals are changed by cold working was presented in Chapter Eight. Here we will discuss, not only the metals, but also the effect of heat treatment on the physical properties of these metals. A general discussion of the heat treatments will be presented first.

### HEAT TREATMENTS

Heat treatments have three functions: they can soften a metal, they can harden a metal, and they can harden the surface of a metal while leaving the interior almost unchanged. These changes are made by producing changes in the metal's structure, grain size, and internal residual stresses. There are four controlling factors in the heat treatment process:

1. The temperature to which the metal is heated.
2. The time that the metal is held at this temperature.
3. The temperature to which the metal is cooled.
4. The rate of cooling.

#### Softening the Metal

Metals are softened to improve the ductility of the metal and thus improve its ability to be formed into desired shapes, and to improve its machinability. A metal loses much of its strength during the softening process so, after forming, it is customary to heat treat the metal to a harder and stronger form. *Annealing* is the general term for the softening process. There are several specific types of annealing, but they all consist of a slow cooling process.

#### Hardening and Tempering

The purpose of a hardening heat treatment is to increase the strength and hardness of the metal. It requires heating the metal to a temperature where certain phase changes occur, called the *transformation temperature,* and then *quenching* (rapidly cooling) it.

There are several quenching media used. Water is the most rapid cooling medium.

Hot oil and molten lead provide higher quench temperatures.
These cool the metal at a slower rate than water. The final temperature is also higher.

*Tempering* is a follow up heat treatment that relieves several problems that result from the quenching process. One of these is the presence of internal residual stresses. These can lead to a problem called *stress corrosion* (see Chapter Fourteen).

A second problem is that the hardness that results from the quench is accompanied by a reduction in ductility. This increased brittleness is undesirable in structural materials. Tempering consists of: reheating the material to a temperature below the transformation range, holding it there for a short time, and then cooling it in air.

## Surface Hardening

The purpose of *surface hardening* is to provide a surface that will withstand wear while retaining the toughness of the core material.

Surface hardening methods can be divided into two groups those that add a substance to the surface and those that use a simple quenching technique.

Most surface hardening methods used in aircraft materials use the additive method. IT consists of heating the metal in the presence of a substance that reacts with, or diffuses into, the surface. A more detailed explanation will be provided when the individual metals are discussed.

The non additive method is essentially a local quenching process. The area to be hardened is heated using a flame or an electrical induction coil and then quenched.

## FERROUS METALS

While most materials used in aircraft construction are non ferrous, aircraft engines, engine mounts, hydraulic lines, control cables and other parts are some alloy of iron. A metallic material that contains at least 50 percent iron is classified as a ferrous metal or alloy. The simplest ferrous metal is plain carbon steel, consisting of less than one percent carbon.

We discussed some physical properties of plain carbon steel in Chapter Eight. We also discussed the effects of cold working in a simple tension test. We learned that the main advantage of cold working was the result of movement of the dislocations during the yielding process. This raised the apparent yield point but did not improve the ultimate strength.

Rolling the material aligns the grains in the direction of rolling. This increases the ultimate tensile strength in the direction of rolling. The rolling is done while the steel is either hot or cold. The cold rolled steels produce the higher strength and hardness, but are lower in ductility.

Low carbon steels (up to 0.30 percent carbon) have low strength and high ductility. They cannot be hardened by heat treatment but can be surface hardened. They can also be hardened by cold working. Annealing will soften cold worked steels.

Medium carbon steels (0.31 to 0.55 percent carbon) can be hardened by heat treatment, but hardening is limited to thin sections. Medium carbon steels in the quenched and tempered condition have a good balance between strength and ductility. They are the most widely used steels for structural materials.

High carbon steels (0.56 to 1 percent carbon) are readily hardened. They are suitable for wear resistant parts, spring steels, ball bearings, and tool steels. They have high strength but lack ductility, thus are not often found in aircraft construction.

## HEAT TREATMENT OF FERROUS METALS

Iron has the property of existing as a solid in two different crystal forms. At temperatures below 1670°F it is in the body centered cubic form and is called *alpha iron.* Above this temperature, called the *upper transformation temperature,* the crystal structure shifts to the face centered cubic form and is called *gamma iron.* The iron becomes non magnetic when this transformation takes place.

Lowering the temperature to the *lower transformation temperature* produces a reverse in the crystal structure from gamma to alpha iron. Carbon has a limited solubility in iron at room temperature of only about 0.005 percent.

Excess carbon can exist in either of two forms. It can form an intermetallic compound, called iron carbide ($Fe_3C$), or it can exist as graphite flakes or spheroids.

*Austenite* is the name given to steel when it is in the gamma iron form. Upon cooling the austenite transforms (decomposes) into one of several other forms depending upon the rate of cooling.

The forms of steel that we will discuss and the processes that create them are called *Martensite, Pearlite* and *Bainite.*

Figure 10.1 is a schematic diagram, called the T T T diagram, which describes the decomposition of austenite in a given steel.

The three Tees represent **Temperature, Time** and **Transformation.**

As the carbon, or other alloying elements, content changes a different diagram results. Thus the diagram shown is for a "typical" medium carbon steel that has an upper transformation temperature of 1450°F.

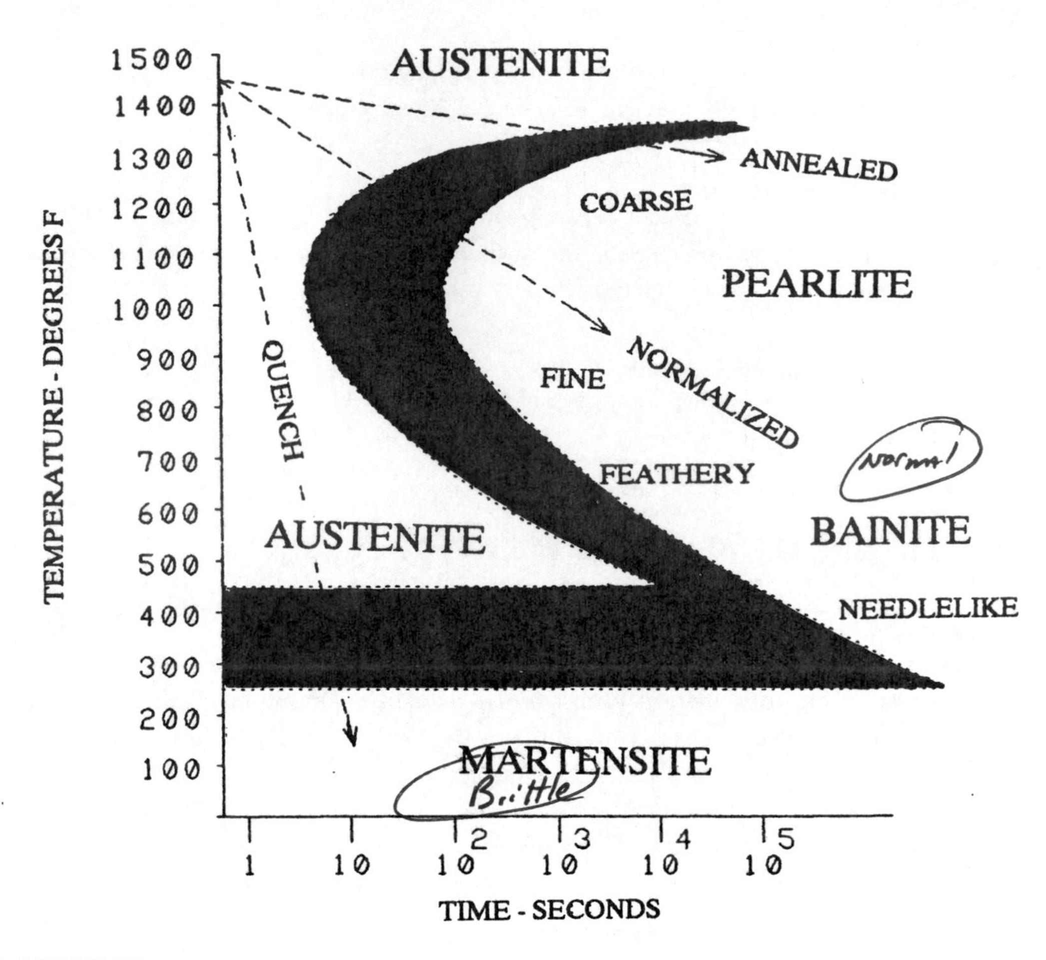


*Figure 10.1 T-T-T Diagram.*

The T T T diagram can only be used as follows:

Start at time = 0 and at temperature above the dark portion (the transformation zone) of the curve. This is at 1450° F in the figure. As the temperature is lowered as a function of time follow one of the dashed lines labeled Annealed, Normalized or Quench.

When the transformation zone is reached the transformation from Gamma iron (face centered cubic) to the Alpha iron (body centered cubic) begins. When the transformation zone is exited the transformation is completed.

Any further entrance into the transformation zone does not effect the composition of the steel.

If it is desired to return the steel to the austenite form, reheating to above the upper transformation temperature is required.

## Softening of Ferrous Metals

Annealing consists of heating the steel to above its upper transformation temperature and then slowly cooling it. This is done by leaving it in the furnace and turning the furnace heat off. This produces a form of the metal called *Pearlite.*

This coarse grained form of steel has low strength, high ductility, and is easily machined. The iron carbide in pearlite is in the form of layers or plates. If the temperature is held at 1330°F for a prolonged period, the iron carbide plates assume a rounded globular form called *Spheroidized steel.*

This is the softest, most ductile form that can be obtained by heat treatment. The spheroidized steel is extremely machinable.

Another softening process is called *Normalizing.* The cooling process consists of removing the steel from the furnace and exposing it to air. This "air quench" produces pearlite, but with a finer grain size than that produced by annealing.

The properties of the normalized steel are; greater strength and reduced ductility than annealed steel. Normalizing is used to refine the structure of steel castings, thus improving their properties.

## Hardening of Ferrous Metals

If the steel is quenched in a molten lead or salt bath at a temperature of between about 500 and 800°F and then held in the bath until the transformation is complete, another form of the steel called *Banite* results. Banite has high strength and reasonable ductility. Choice of the bath temperature will vary these properties.

Quenching the austenitic steel in cold water is the most severe heat treatment. The resulting form of the steel is called *Martensite.* It is a supersaturated solid solution of carbon in iron. Martensite is a very brittle material with high tensile strength and hardness.

The quench produces high internal stresses in the metal and distortion of thin parts results. The internal stresses and the lack of ductility are unacceptable in structural applications, thus a follow up tempering is required.

Tempering (or Stress Relieving) consists of reheating the quenched steel to about 1000°F and holding it there for about one hour. This increases the ductility of the steel but the tensile strength is reduced.

The resulting steel is called *Tempered Martensite.*

## Surface Hardening of Ferrous Metals

The three most common methods of hardening the surface of a metal by atomic diffusion of another substance are: carburizing, nitriding and cyaniding.

*Carburizing* produces a high-carbon surface on a low-carbon steel.

Low-carbon steel is heated to about 1650°F in the presence of solid, gaseous, or liquid substances containing carbon. The carbon enters the surface of the low-carbon steel and is retained by quenching. The hardened surface is about .05 inches deep. Larger gears, camshafts and bearings are typical parts that are surface hardened by this method. The quench may produce warping of the parts, so thin sections are hardened by a less severe process.

*Nitriding* is the process used on hardened and tempered alloy steels.

The process consists of heating the steel to about 950°F in an atmosphere of ammonia gas for about 50 hours. No further heat treatment is required. This treatment produces a hardened surface of only about one third the thickness of the carburizing treatment. The advantages of this method are: extreme hardness, resistance to reheating to 800°F without softening, and improved corrosion resistance. Warping of thin parts will not occur as a quench is not used.

*Cyaniding* is a process used on low-carbon steel.

The process consists of heating low-carbon steel parts in a molten sodium cyanide bath at 1600°F for about one hour followed by an oil or water quench. Again the quench may produce distortion of thin parts.
The hardness is about the same as carburizing. Parts that are hardened by this method include: screws, nuts and bolts, and small gears. The hardened surface thickness is only about one-fourth that of carburizing.

## NONFERROUS METALS

Aluminum, magnesium, titanium, and beryllium are classified as light metals because their density is much less than that of steel. This is a definite asset in aircraft construction.

## ALUMINUM

Pure aluminum has low strength and is not suitable as a structural material. But it is used as a corrosion resisting material in the clad alloys. There are two main classes of aluminum alloys: Wrought alloys and Cast alloys.
Most aircraft parts are wrought alloys, which we will discuss further.

## Wrought Aluminum Alloys

The metals that are used as the principal alloying elements in wrought aluminum alloys are: manganese, silicon, magnesium, and zinc. There are two similar numbering systems (one for wrought and one for cast) aluminum alloys. The wrought series and the major alloying elements are as follows:

**Designations for Wrought Aluminum Alloy Groups**

| Major Alloying Element | Alloy No. |
|---|---|
| Aluminum-99%+ pure | 1xxx |
| Copper | 2xxx |
| Manganese | 3xxx |
| Silicon | 4xxx |
| Magnesium | 5xxx |
| Magnesium and Silicon | 6xxx |
| Zinc | 7xxx |
| Other elements | 8xxx |
| Unused series | 9xxx |

For aluminum alloys in the 2xxx through 8xxx groups the second digit shows modifications to the original alloy. If no modification exists the second digit will be zero. The third and fourth digits serve only to identify different alloys within the group.

The four number group is followed by a temper designation consisting of a letter and one or more numbers. There are over one hundred wrought alloys in use today but aircraft construction materials consist primarily of two or three of these. We will investigate these further.

## 2000 Series

Copper is the principal alloying element in this group. These alloys require heat-treatment to get optimum properties. After heat-treatment the mechanical properties of these alloys are comparable to mild steel. The alloys of this group do not have good corrosion resistance and must be protected by cladding with an alloy that is corrosion resistant. The most widely used aluminum alloy of this group in aircraft construction is 2024. We will discuss this further in Chapter Fourteen.

### 7000 Series

Zinc is the major alloying element in this group. If small percentages of magnesium, copper and chromium are coupled with the zinc, the result is a heat-treatable aluminum alloy of great strength. The foremost member of this group is alloy 7075, which is among the strongest alloys available and is used in air-frame structures and highly stressed parts. Unfortunately this alloy is highly susceptible to stress corrosion.

Work hardening (cold working) increases the strength of wrought aluminum alloys. Work hardening is done by rolling, bending, drawing, forging, or extruding. These processes must be done at temperatures below the recrystallization temperature (650-800°F).

## HEAT TREATMENT OF WROUGHT ALUMINUM ALLOYS

### Softening of Wrought Aluminum Alloys

Aluminum alloys may be annealed to remove prior hardening or strengthening effects of cold working or heat treatment.

This consists of heating the alloy to above its recrystallization temperature of 650-800°F (depending upon the alloy) and holding it at this temperature until the new crystals are formed. This process is used for previously work-hardened alloys. For alloys that have been hardened by heat treatment the process is done to precipitate the alloying particles out of solution. The rate of cooling is important.

If the alloy is work-hardened during fabrication it loses its ductility and becomes brittle. Annealing restores the ductility so that it can be further shaped. In extensive shaping, several successive annealing processes may be required.

### Hardening of Wrought Aluminum Alloys

*Solid-Solution Hardening.*

Alloys are made by dissolving other metals in aluminum to form solid solutions. In our discussion of crystal imperfections in Chapter Four, we learned that substitutions and interstitial atoms exist when foreign atoms are present. Here, atoms of the alloying metals replace certain aluminum atoms and/or form interstitial atoms. In either case, the metallurgical structure is distorted by the new atoms in the structure, thus increasing the strength.

These alloys can be further strengthened by heat-treatments and/or work hardening.

*Precipitation Hardening.*

The alloying elements in heat-treatable aluminum alloys are more soluble at higher temperatures than they are at room temperature. When these alloys are heated and then quenched, a supersaturated condition is produced. The alloy does not attain its full strength until the alloying elements precipitate out of solution after a length of time. This effect is called *precipitation* or *age hardening.*

Artificially aging by reheating the alloy is used when maximum strength is desired in a short time span. The alloy loses ductility when strength is increased in age hardening. This is often undesirable when forming has not been completed. Rivets, for instance, can not be driven in the fully hardened condition.

They will split because they have lost their ductility. Refrigeration will prevent age hardening, so rivets are stored in dry ice until they are installed.

Stress-relieving is done to eliminate internal stresses resulting from a cold quench. This is similar to the tempering process of ferrous materials. Some strength is lost as the ductility is increased during the process.

## Temper Designations for Heat Treatable Alloys

Aircraft aluminum alloys including 2014, 2024, 7075, and 7178 are heat treatable. The alloy number is followed by the temper process. Some more common designations are:

O-Fully annealed
H-Strain hardened
T3-solution heat treated then cold worked
T4-solution heat treated and naturally aged
T6-solution heat treated then artificially aged

## MAGNESIUM

Magnesium is the lightest of structural metals. It is not used in its pure state, but is alloyed with other metals such as aluminum, zinc and manganese. Magnesium alloys are somewhat low in strength, but their low density gives them a strength to weight ratio comparable with high-strength aluminum alloys.

Magnesium is the easiest of structural materials to machine. Magnesium alloys have poor corrosion resistance when subjected to marine environments. The alloys are highly anodic and precautions must be taken to isolate them from other metals or galvanic corrosion will result.

Magnesium has the hexagonal close-packed crystal structure which makes its room temperature ductility poor. At higher temperatures the ductility improves. Forming is done in the hot condition. Casting properties of magnesium are excellent, thus many aircraft castings are made of magnesium alloys.

Heat treatment of cast magnesium alloy involves precipitation hardening. Aging is done by heating to 430°F for 5 hours. If wrought magnesium alloy sheet is to be formed into parts it is supplied in the annealed condition and hardened by cold working.

## TITANIUM

Titanium and its alloys are about 40 percent lighter than steel and 60 percent heavier than aluminum. They have a strength to weight ratio that is higher than any structural metal. This ratio exists from -420°F up to +1000°F.

The second outstanding property of titanium is its corrosion resistance.

Titanium is difficult to fabricate because it is susceptible to hydrogen, oxygen and nitrogen impurities that cause the metal to become brittle. High temperature processing, including welding must be done in a protective atmosphere to prevent diffusion of these gasses into the metal.

Titanium is one of the few *allotropic* metals (steel is another). This means that its crystalline form changes when heated.

At room temperature it has the close-packed hexagonal structure. This is called the *alpha phase*. Upon heating to 1625°F, the alpha phase transforms into a body centered cubic form, known as the *beta phase*.

Alloying elements stabilize one or the other of the two phases. Aluminum and carbon, for instance, stabilize the alpha phase by raising the transformation temperature.

Copper, chromium, iron and molybdenum are beta stabilizers. They lower the transformation temperature, allowing the beta phase to be stable at lower temperatures, even at room temperature.

The physical properties of titanium are affected by the allotropic phases. The beta phase is much stronger and more brittle than the alpha phase.

Titanium alloys are classified into three groups according to their allotropic phases: the alpha, beta and alpha-beta alloys.

The alpha phase alloys are the most widely used in aircraft and space applications. The most widely used alpha alloy is a 5 percent aluminum and 2.5 percent tin alloy designated as Ti-5Al-2.5Sn.

It has a tensile strength of 125,000 psi at room temperature, 18 percent ductility, a modulus of elasticity of 16,000,000 psi, and retains its strength at temperatures up to 1000°F.

## BERYLLIUM

Beryllium is a metal that has exceptional properties and serious deficiencies. the advantages include light weight (one third lighter than aluminum), stiffness to weight ratio (six times that of high strength steels), excellent thermal conductivity, and excellent electrical conductivity.

Beryllium has the hexagonal close-packed crystal structure and is highly sensitive to impurities which combine to make it very brittle at room temperatures. another limitation is that it is very toxic if inhaled or ingested.

To overcome brittleness problems, beryllium is most useful as an alloying element and as a composite in matrices of titanium and aluminum. Beryllium-wire-reinforced aluminum sheet is used to make pressure bottles. Another useful product is beryllium-reinforced titanium alloy composites with a tensile strength of 140,000 psi.

## REVIEW PROBLEMS

1. Heat treatment of a metal consists of heating the metal to some "critical" temperature, then:

   (a) adding alloying metals.
   (b) tempering the metal.
   (c) accurately controlling the rate of cooling.
   (d) quenching the metal.

2. Quenching of a carbon steel produces "Martensite" that is:

   (a) a strong hard steel that is a good structural metal.
   (b) easily machined.
   (c) not ductile enough for structural use.
   (d) free from residual stress.

3. Spheroidizing of steel is done to:

   (a) increase its strength.
   (b) increase its machinability.
   (c) increase its hardness.
   (d) reduce internal stresses.

4. Which of these surface hardening methods produces the thickest hardened surface?

(a) Carburizing
(b) Nitriding
(c) Cyaniding

5. Aluminum alloys attain their strength by:

(a) quenching.
(b) cold working.
(c) precipitation.
(d) all of the above.

6. Which of the following has the least corrosion resistance?

(a) Aluminum
(b) Magnesium
(c) Titanium
(d) Beryllium

7. Which of the following is the most toxic to personnel?

(a) Aluminum
(b) Magnesium
(c) Titanium
(d) Beryllium

# CHAPTER ELEVEN

# COMPOSITE MATERIALS

## INTRODUCTION

The word "composite" in composite materials means that two or more materials are combined to form another new useful material. Unlike alloying, in which the combining is done on a microscopic scale, composite materials are combined on a macroscopic scale.

Several properties of the individual materials can be improved by the forming of them into a composite material. These include the properties that were discussed in Chapter One:

Static Strength
Rigidity
Service Life (fatigue life and corrosion resistance)

In addition, other improvements include:

Wear Resistance
Weight Reduction
Thermal Insulation
Acoustical Insulation

Composite materials have been in use for a long time. History recounts that the ancient Israelites used straw to add to the strength of mud bricks.

The Egyptians were the first to use plywood. In 1850 a French gardener discovered that steel rods and wire mesh embedded in concrete improved its strength.

Fiberglass came into general use in the 1950s to make strong, light panels for sport cars and rust-resistant boats. It is made of several layers of woven glass cloth. The layers are placed in a mold, one at a time, and are impregnated with an epoxy resin.

The pliable layers and wet resin are then pressed together and heated. When this "curing" process is completed the material is rigid and can be cut or drilled as desired. Fiberglass is easy to work with and is somewhat inexpensive, thus is a good composite for "homebuilt" or "kit" light aircraft.

Plywood is a laminated composite application that makes for a very strong lightweight structure. The British Mosquito, shown in Figure 11.1, was a World War II high performance fighter made entirely of plywood, as was Howard Hughes' famous Spruce Goose.

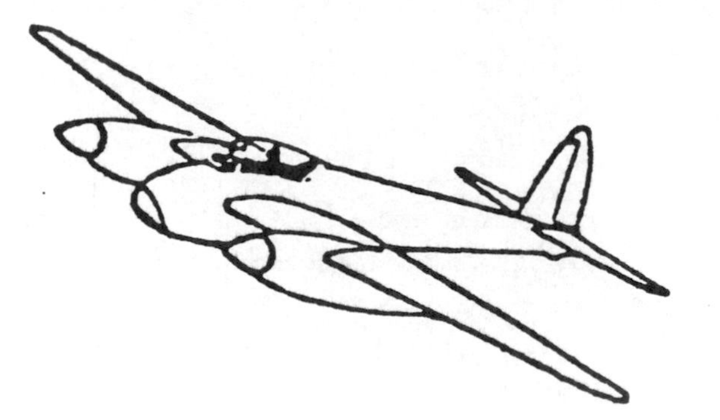

*Figure 11.1 Mosquito Fighter.*

There are many other composites that have superior strength and other desirable qualities. The bad news is that they are much more expensive and are more difficult to manufacture.

Military aircraft, which rely on light weight to attain superior performance, have been the testing grounds for these "advanced composites." The aircraft industry has taken a step - by - step evolution process in the use of composites.

At first, they were used in secondary structures, which would not cause catastrophic results if they failed in flight, such as filets, fairings, wheel doors etc.

As confidence grew, and improvements in the manufacturing processes were made, more critical parts of aircraft such as wings, control surfaces, fuselage and empennages were made of composite materials.

Composite materials were developed because no single structural material could be found that had all the physical properties that were desired. The aerospace industry was one of the major forces behind the development program. Aluminum alloys have high strength and stiffness and low weight, but both corrosion and fatigue damage have caused many problems in aircraft structures.

The first composites developed during World War II were fiberglass-reinforced plastics. These were inexpensive and are still widely used today in many applications such as boats and sport car bodies. More advanced fiber-reinforced composites have been restricted in their commercial use because of their high cost, lack of raw material sources and efficient manufacturing methods. The military and aerospace programs have done much to further the development of these composites and thus reduce the cost. The airlines have become major users of fiber-reinforced composites because of weight reduction considerations and resulting fuel conservation.

## CLASSIFICATION OF COMPOSITE MATERIALS

## PARTICULATE-REINFORCED COMPOSITES

*Particulate-reinforced* composites consist of particles of one material embedded in the matrix of another material.

Concrete is the most common example of a particulate. Concrete is made by mixing particles of sand, gravel, cement and water. The cement and water chemically react and harden and bind the sand and gravel together.

One type of particulate composite is made by suspending flakes of nonmetallic materials such as mica or glass in a matrix of glass or plastic.

The flakes are two-dimensional, they have strength and stiffness in two dimensions, whereas fibers of the same material would have these properties in only one dimension.

Other particulates use metallic flakes in a nonmetallic matrix. An example of this type of composite is found in solid rocket fuels, some of which use aluminum powder and oxidizers in a polyurethane matrix. Aluminum paint, which suspends aluminum flakes in paint, is another common example.

Another type of particulate composite uses embedded metallic particles in a metallic matrix. Unlike an alloy, the particles do not dissolve.

Some metals, such as chromium, molybdenum, and tungsten are brittle at room temperature and must be heated to allow machining.

But when they are in particle form they can be suspended in a ductile metallic matrix. The resulting composite is ductile, yet maintains the properties of the brittle metals at high temperatures.

The final class of particulate composites are called *cermets*. There are two classes of cermets; *oxide-based* and *carbide-based*.

Oxide-based cermets can be either oxide particles in a metal matrix or vice versa. These composites are used in tool making and in high temperature environments where erosion resistance is desired.

Carbide-based cermets have particles of chromium, titanium, and tungsten in a matrix of cobalt or nickel. Tungsten carbide in a cobalt matrix has a high degree of hardness and is used in parts such as wire-drawing dies.

Chromium carbide in a cobalt matrix has excellent corrosion and abrasion resistance and is used in valves. Titanium carbide in a nickel matrix has good high temperature properties and is used in gas turbine parts.

## FIBER-REINFORCED COMPOSITES

In this type of composite the reinforcing materials have lengths much greater than their widths. These reinforcing materials are called fibers. Long fibers are much stiffer and stronger than the same material in bulk form. For instance, ordinary plate glass has an ultimate tensile strength of only a few thousand pounds per square inch, but commercially available glass fibers have strengths of up to 700,000 psi. This improvement in strength is due to the nearly perfect structure of the fiber. The crystals of the fiber are aligned along the fiber axis and there are fewer dislocations than in the bulk form.

Two categories of fiber-reinforced composites exist. These are the *discontinuous fiber* (short fiber) and the *continuous fiber* (long fiber). If the properties of the composite vary with the length of the fiber, it is in the discontinuous category. If the properties remain the same with fiber length, it is a continuous fiber.
Most continuous fiber reinforced composites contain fibers that run the entire length of the composite part.

## LAMINAR COMPOSITES

A single layer of a fiber-reinforced composite is called a *lamina*. When layers of lamina are assembled as shown in Figure 11.2, the resultant material is called a *laminate* or, more popularly, a *laminar composite*.

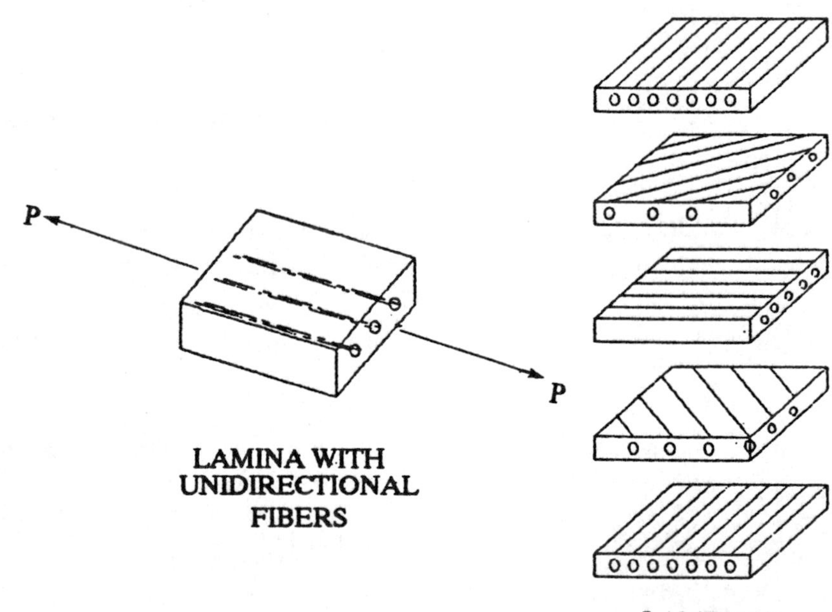


*Figure 11.2 Lamina and laminate.*

One major advantage of laminar composites is the ability to tailor the direction of the reinforcing fibers to provide strength and stiffness as desired. For instance, six layers of a nine-ply laminate can be oriented in one direction and the other three at 90° to that direction. The resulting laminate has about 50 percent more tensile strength in the direction of the six layer fibers than in the three layer fibers. Stiffness is also greater in one direction than in the other. These properties are also influenced by the order in which the laminae are arranged.

## Reinforcing Fibers

There are many advantages of fiber composites over bulk materials. Nearly all non-composite materials used in structural applications fail because of flaws that grow until failure occurs. Flaws in fibers are limited to the diameter of the fiber. In addition, failure of one fiber is not propagated to the surrounding fibers.

The matrix transmits the stress normally carried by the broken fiber to the surrounding fibers as shown in Figure 11.3(a).

The tension stress in the broken center fiber, $f_t$, is transmitted to the surrounding fibers by shear stresses, $f_s$, in the matrix. This is shown in the inset, Figure 11.3(b).

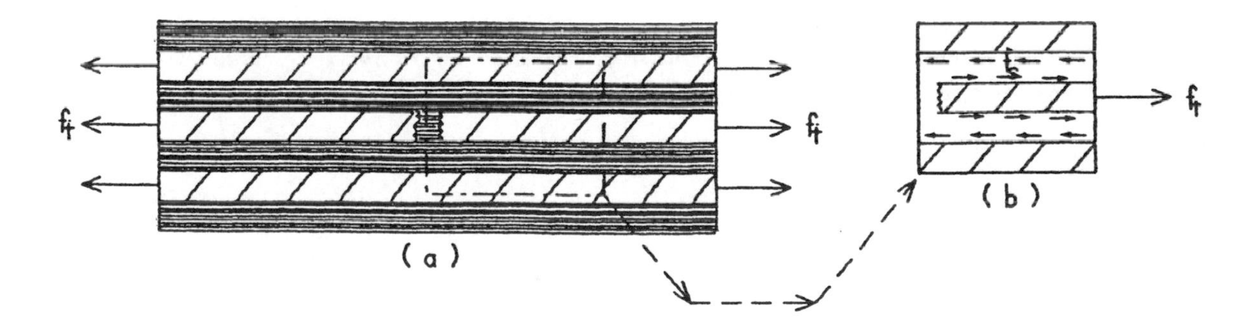

*Figure 11.3 Schematic of broken fiber load transmission.*

A nylon fiber called Kevlar is also being used extensively. Kevlar offers great durability and is often used as a "crack stopper" and to absorb impact loads. Kevlar laminae are interlayered with those of carbon for a

### *Composite Materials*

A new structural composite called Arall has recently been developed. Arall is the acronym for aramid aluminum laminate. Arall 1 is but one of the many possible combinations of aluminum and aramid fibers.

It is a laminate of 0.012 inch thick 7075-T6 aluminum sheet and aramid pre-preg of 0.008 inches thick in an epoxy matrix. The laminate has three layers of aluminum with two layers of aramid.

The final product is 0.052 inches thick with aluminum sheet on the top and bottom. Primary application for Arall 1 is in structural areas where tension is the primary stress and where fatigue and fracture are critical. One of Arall's strong points is its resistance to fatigue crack growth.

Among the advantages of the mechanical properties of the fibrous composites is an extremely long fatigue life. On some occasions the testing machines failed before the fiber part failed in fatigue. The fatigue life of most composites will be several times the expected life of the airplane.

Weight saving is the driving force for composite technology. For most individual articles such as a rudder, manufacturers get from 15 to 30 percent weight reduction by using composite materials.

Some predictions indicate that for vehicles where significant amounts of composites are used that a total weight reduction of from 25 to 50 percent may be gotten with manufacturing cost reductions of 25 to 35 percent.

## Quality Control

Strict quality control is required in the manufacture and fabrication of composite materials as there may be many types of material defects that are not found in metals.

Among the common defects that have to be controlled are:

inter-laminar voids due to air entrapment,
delamination,
insufficient resin,
incomplete curing of the resin,
excess resin between the layers,
excess matrix voids in porosity,
incorrect orientation of laminae material directions,
damaged fibers,
wrinkles or ridges caused by improper lay-up,
inclusion of foreign matter,
unacceptable joints in layers,
and variations in thickness.

## Basic Failure Modes

Because continuous fiber composites are heterogeneous and metals are homogeneous, the fractures of composites are very different from those of metals. In metals there is a single propagating crack while in composites the damage area is caused by matrix cracking, fiber breaking and delamination. These combine to result in failure of the part.

Loads that put the fibers in tension and straighten them are most easily handled by these composites. In fatigue, tension loaded composites are superior to nearly any other material of similar weight. Yet in compression the fibers in the laminates have a tendency to bend or buckle. The fibers are not perfectly straight and axial compression produces shear between the fibers and the matrix. This shear produces component stresses of tension in the matrix, as was discussed in Chapter Five, that may cause matrix failure. This type of failure is described as an "in-plane" failure and some of these are shown in Figure 11.4.

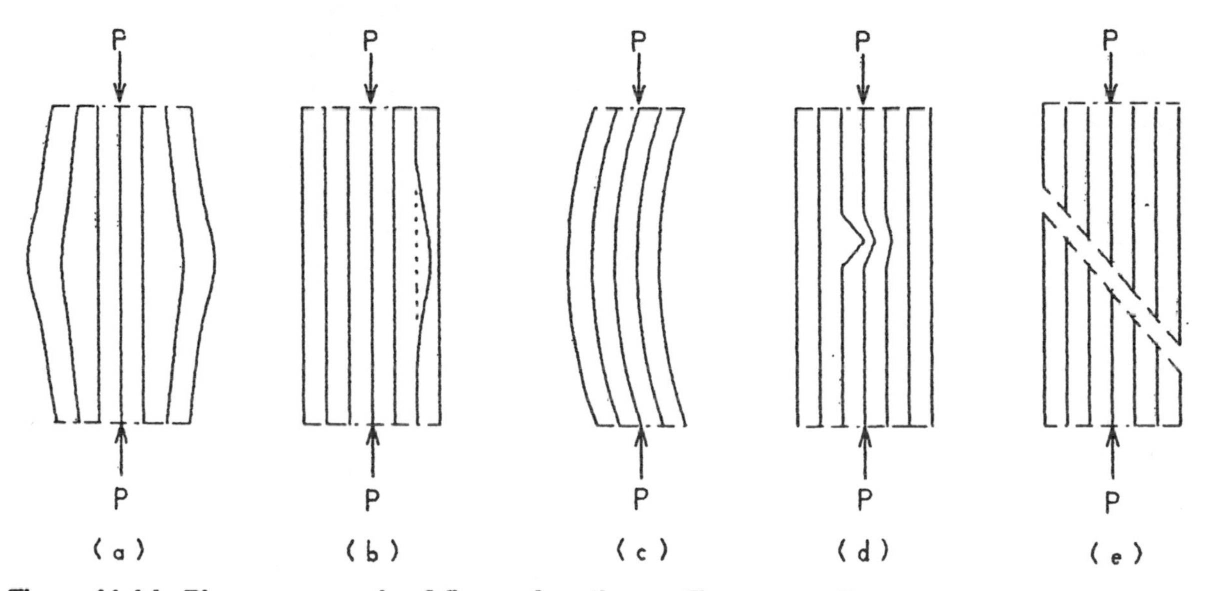


***Figure 11.4** In-Plane compression failures of continuous fiber composites.*

Poisson's ratio, discussed in Chapter Eight, describes how the material becomes wider when placed in compression. This results in the development of lateral tension stresses and may cause matrix tension failure as shown in Figure 11.4(a). If a fiber buckles, the fiber-matrix interface may fail in shear as shown in Figure 11.4(b). If the matrix is ductile and the interface is strong, but this may not happen and the fiber can fail by bending as shown in Figure 11.4(c). Shear crippling is a microscopic failure and is the result of kink-band formation. An attempt to show this on a greatly magnified scale is shown in Figure 11.4(d). The final failure, shown in Figure 11.4(e), is the 45° shear failure of the fibers that results from compression loads that was discussed in Chapter Five.

Failures between the laminates are called Out-of-Plane Delamination Failures. Discussion of these is complicated and is beyond the scope of this book.

## AIRCRAFT ACCIDENT INVESTIGATIONS

Determination of the causes of failures of composites in the field is going to be a very difficult problem for a long time to come. Laboratory analysis has improved greatly in the past few years, but there is no way to apply this ability to field analysis to date.

Some of the many factors that complicate the analysis are: the type of material, the orientation of the laminae, the processing of the composite, the moisture content, the loading type and direction, etc.

An impact load makes a composite act in a brittle fashion and it shatters into small pieces. When loaded in compression the ends become frayed. The name given the appearance is "brooming" as the ends resemble the end of a straw broom.
Both static compression and fatigue failures appear the same.

As was discussed earlier and illustrated in Figure 11.4, several different types of failure can result from compression failures. The best approach to take in an investigation involving composite materials is to seek laboratory analysis by experts in composite materials.

### Post Crash Fires

As both the graphite and the matrix of graphite/epoxy fiber composites are made from petroleum, it is not surprising that fires involving this material are common.

In a crash and subsequent fire, the heat of the fire will release the carbon fibers as ultra-fine whiskers that are carried up on the heat thermals and fall out perhaps miles from the crash site.

These whiskers are electrically conductive. For a while there was quite a scare about the potential disastrous nature of these carbon fibers settling upon a community and causing all the transformers on the power poles to explode and bring all the machinery to a stop. It still is a problem but not of the magnitude that was first expected. Fire fighters should have information on how to spray the wreckage to contain these fibers. That will complicate the investigation unless the right material is sprayed upon the fire.

These fibers must be contained because they could disrupt the communications for the area of the crash site. Carbon whiskers can be very disastrous for computers and other electronics that use micro-circuit technology in that they can settle across a circuit and short it out. The process will vaporize the fiber but meanwhile, the harm has been done.

Other problems concerning composite fires include the inability to identify inflight fires from fires on the ground, the destruction of structural evidence by the fire, and because the composite will continue to smolder long after the flames of the fire have been extinguished.

## OTHER APPLICATIONS

The tailoring of fibrous laminates has already led to some very interesting and advantageous applications to aircraft technology.

We have noticed that several proposals for highly maneuverable fighters will have forward swept wings and this is made possible by the aeroelastic tailoring developments.

The forward swept wing (FSW) is just as good for delaying shock effects and Mach effects as is the aft swept wing. Wings have been swept aft traditionally because they are less susceptible to flutter and divergence.

In 1973 Col. Norris Krone, Jr. published a professional paper, "Divergence Elimination with Advanced Composites." Krone showed mathematically that it was possible to build the FSW aircraft without the wings tearing themselves off.

Krone's research showed that the desired torsional stiffness can be tailored by fiber orientation to resist torsional flexing that would lead to flutter and divergence. Therefore we can design the forward swept wing fighter.

There are many advantages of maneuverability, CG control, area rule, eliminating wing tip stall, and so forth for the forward swept wing. This ability to control the stiffness of the wing also will allow one to tailor combinations of bending and torsion modes.

For example, during a high-G turn, the wing tips will tend to bend up but the laminae can be laid-up so that when the wing tips flex up they will automatically twist down in torsion. This will decrease the angle of attack and unload the outboard section of the wing.

The load is thereby shifted more toward the fuselage decreasing the bending moment and thus the structure can withstand much higher G turns.

This same tailoring is also in use for helicopter rotor blades. They undergo a very large amount of flexing in bending and in torsion, especially at high forward velocities. This is due to the great difference in relative velocity of the advancing and retreating blades.

One blade is approaching Mach shock effects and the other is on the verge of aerodynamic low speed stall. The aeroelastic tailoring of the flex on these blades has been invaluable.

Additionally, because we need not be concerned about fatigue of fiber composite blades, is a very great advantage. Corrosion resistance of composites is also a plus factor. The resultant blade should be at least 20 percent lighter in weight than one made by conventional metal fabrication methods.

## REVIEW PROBLEMS

1. Concrete is an example of a ________ composite.

(a) fiber-reinforced
(b) laminar
(c) particulate-reinforced

2. Name two aircraft that were made before 1950 of composite materials .
Mosquito, Spruce Goose

3. Composites make swept forward wings practical because:

(a) they can be tailored to resist twisting moments.
(b) they resist fatigue better than metals.
(c) they have better static strength than metals.
(d) they have a higher modulus of elasticity than steel.

4. Arall is a composite which:

(a) is a fiber-reinforced composite.
(b) is a particulate-reinforced.
(c) combines metallic and nonmetallic materials in a laminated composite.

5. In a fiber-reinforced composite the major load carrying component is:

(a) the fiber.
(b) the matrix.
(c) both carry the load equally.

6. True-False Fiber-reinforced composites have poor fatigue life.

7. Fiber-reinforced composites are weakest in:

(a) tension.
(b) compression.
(c) fatigue.

8. True-False. A competent accident investigator can determine the cause of failure of a composite part at the
e.

# CHAPTER TWELVE

# STRESS CONCENTRATIONS

## INTRODUCTION

In Chapter Eleven we learned that composite materials had excellent fatigue resisting qualities. Unfortunately, this is not true for metallic aircraft parts.

Fatigue has been a troublesome cause of reduced service life for many years and, since we continue to make airplanes out of metal, it appears that the problem will persist.

All fatigue failures are the result of fatigue cracks. These cracks nearly always start from regions of high local stress called *stress concentrations* or stress raisers. There are many sources of stress concentrations.

Some stress concentrations may be inside the metal itself, such as non-metallic inclusions (slag or other impurities), and cracks caused by dislocation movement during fabrication. But most stress concentrations are on the surface of the material. Many stress concentrations result from poor design. These include: holes, notches, grooves, sharp corners, threads, etc. We will investigate these in more detail in this chapter.

Other stress concentrations are produced after the airplane is manufactured and placed in service. These are: tool marks, scratches, stone bruises, etc. Personnel in the operations and maintenance field should be aware of the dangers of creating stress raisers that could lead to fatigue failures.

## Lines of Constant Force

The effects of a stress concentration can be best understood by applying the "streamline" analogy from aerodynamics. Consider a cylinder in a smoke wind tunnel such as shown in Figure 12.1(a). The smoke lines are closer together as they pass over and under the cylinder. This shows that the velocity of the air is increasing as it flows around the cylinder.

The analogy is that the force lines in a metal plate with a hole in it and under tension loading are similar to the smoke lines in the wind tunnel. Each line represents the path over which a unit of constant force is transmitted.

Near the hole where the lines draw closer together, an increase in force per unit area, or a stress concentration is shown.

The closer the relative spacing of the lines, the greater the stress concentration. Similar stress concentration analogies are shown for a sharp notch in Figure (12.1(b)). A sudden change in section using a small fillet is shown in Figure (12.1(c)), and the reduced stress concentration resulting from using a large fillet is shown in Figure (12.1(d)).

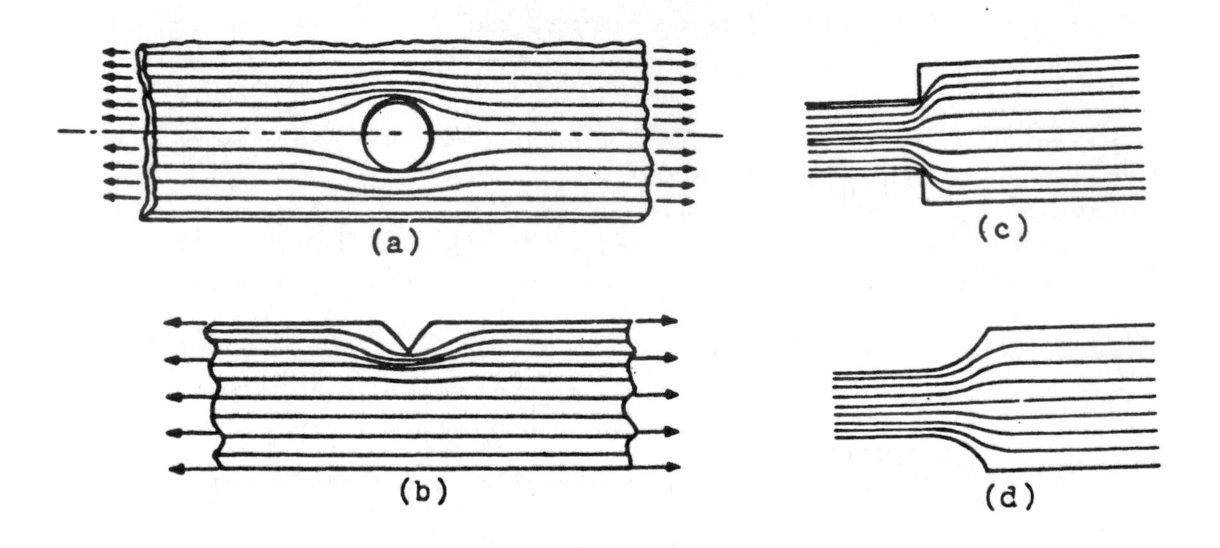


*Figure 12.1 Stress concentration lines of constant force.*

## ELASTIC STRESS CONCENTRATION FACTORS

Increases in local stress caused by stress concentrations are described in terms of a *stress concentration factor*, K, defined as follows:

$$K = \frac{\textit{maximum actual stress}}{\textit{nominal stress}} \qquad (12.1)$$

In a thin plate of metal under tension load, for instance, the nominal tension stress $f_n = P/A$.

A round hole in a wide plate produces a maximum local tension stress at the sides of the hole of three times the nominal stress. This is shown in Figure 12.2(a). The value of the stress concentration factor, K, is 3.

The maximum stress at the edge of the hole decreases as the distance from the side of the hole increases.

This is shown in the stress distribution curve (Figure 12.2(b)).

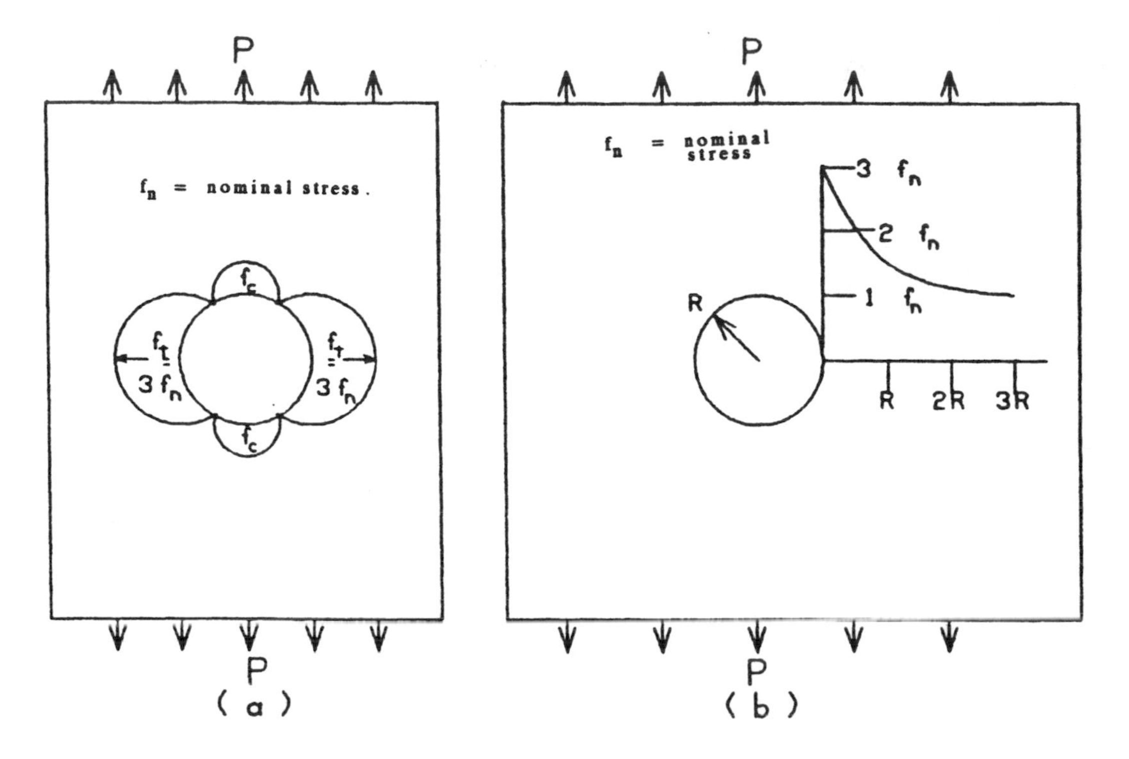


*Figure 12.2 Stress concentration factors about circular hole in a wide plate.*

If a material is in the elastic range, the value of the stress concentration factor is constant. Once the yield point is reached, ductile metals begin the "slip" process and the rate of strain increases.

The rate of stress is sharply reduced at this point and the stress concentration factor is reduced in value. As we discussed in Chapter Four, slip lines result from the movement of dislocations in crystalline metals.

Slip lines unite with other slip lines until slip planes form. Slip planes are cracks and under repeated tensile stress these cracks grow as fatigue cracks.

Ductile ferrous metals have stress-strain diagrams similar to Figure 12.3(a). If the metal shape represented by Figure 12.3(a) has a stress concentration factor $K = 3$, it will reach its yield stress when the nominal tension stress is 10,000 psi.

But, the value of K is reduced once the yield stress is reached because the ductile metal will flow around the stress concentration when the yield stress is reached. Thus a nominal stress of 20,000 psi will not create 60,000 psi stress and cause failure at the stress raiser.

The stress concentration started the fatigue crack in the ductile material but had little influence on the static strength of the material.

## *Stress Concentrations*

To find the effect of a stress concentration on a brittle material, consider a ceramic material.

The stress-strain diagram for this material is shown in Figure 12.3(b). The material is elastic until it fractures. There is no yield point and slip does not occur. The stress concentration factor remains at a constant value until fracture occurs.

Brittle materials are called *notch-sensitive* and stress concentrations will often lead to static failures not to fatigue failures.

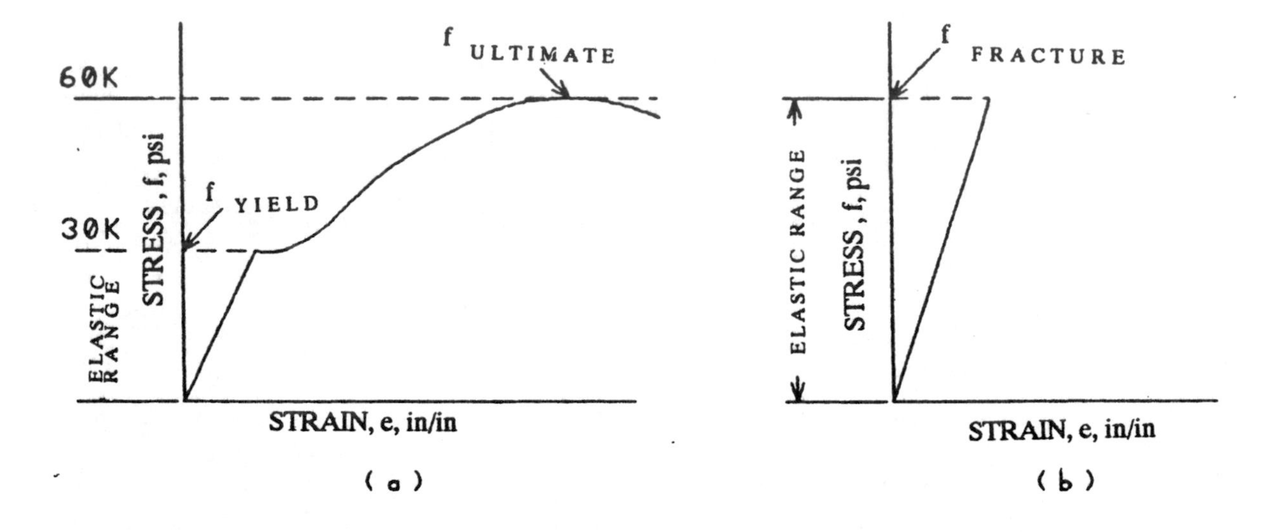


***Figure 12.3 Stress-strain diagram for: (a) ductile ferrous metal; (b) brittle ceramic material.***

## Values of Elastic Stress Concentration Factors

The analytical calculations for all stress concentration factors are difficult. Experimental values can be found by photo-elastic methods.

One method is to make a duplicate of the part out of plastic and to apply loads with similar orientation to those expected in use. When the plastic part is viewed under polarized light colored lines appear. These show the intensity and direction of the principal stresses.

Another method is to coat the part with a brittle material and then load it. The brittle material will crack when the load is applied and the strain reaches a certain value. The direction and number of cracks show the principal tension stress.

Electrical strain gages can be applied to large stress concentrations to find the magnitude and direction of tension stresses. These and other methods are available commercially.

Both mathematical and experimental analysis of the most common stress raisers, such as holes and notches have been made with close correlation between the two methods. Calculation of stress concentration factors in round and elliptical holes in an infinitely wide plate are found by:

$$K = 1 + 2\left(\frac{a}{b}\right) \tag{12.2}$$

Where: a = width of hole
b = height of hole (see Figure 12.4)

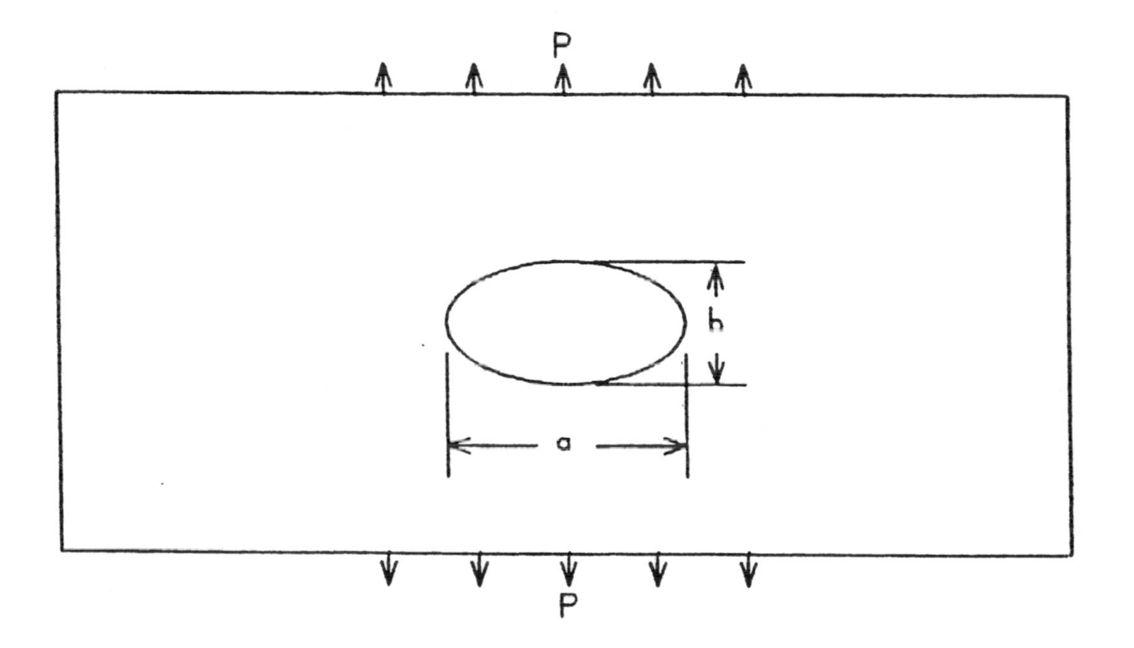


***Figure 12.4 Stress concentration factors for elliptical holes in an infinitely wide plate.***

Applying equation 12.2 to an elliptical hole with a = 2b, we find the value of K = 5, and if the hole has the dimensions of b = 2a, we find K = 2. These are shown in Figure 12.5.

Equation 12.2 can be applied to a circular hole where a = b, so K = 3.

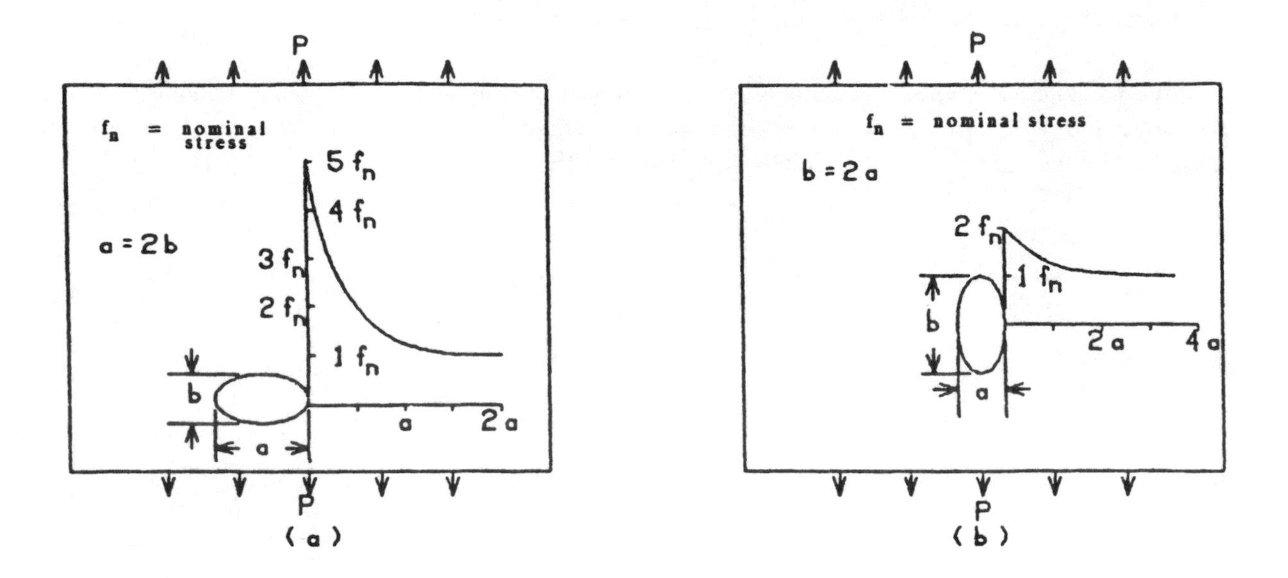


*Figure 12.5 Stress concentration factors and stress distribution for elliptical holes.*

An extreme case involving a crack perpendicular to the tension load (b = 0) shows a/b has the value of infinity, so K = ∞. Thus cracks spread in the direction 90° to the applied tension load.

A maintenance practice known as "stop drilling" has been used for many years. It consists of drilling a small hole at the apparent end of a crack. The object is to stop the crack from spreading.

The crack has microscopically progressed some distance beyond the visible end of the crack so unlikely the exact end can be located. If, by some chance, the exact end of the crack is found, the drilling will reduce the K factor from ∞ to 3.

This may slow the spreading of the crack, but eventually it will continue.

Stop drilling is not a satisfactory way to stop cracks from spreading.

Finally, consider a crack parallel to the tension stress, (a = 0). The value of a/b is now 0 and K = 1. No stress raiser is present with this orientation of the crack.

If a mechanic removes a stone bruise on a propeller, he should always file, grind, or sand in the direction of the tension load (along the length of the blade). The K values for the proceeding cases are shown in Figure 12.6.

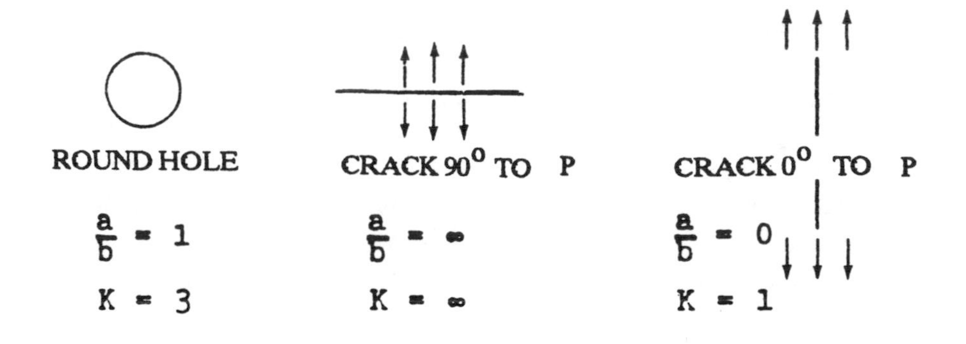


*Figure 12.6 Stress concentration factors for a round hole and cracks in a wide plate.*

Bolted or pin loaded holes are those in which the load is transmitted to the plate at a single point, not uniformly along the entire plate. This results in higher stress concentration factors than for open holes. The values vary as the hole diameter/plate width ratio changes. For a 1/5 ratio, for instance, the value of K = 6.55. A bolt or pin loaded hole is shown in Figure 12.7.

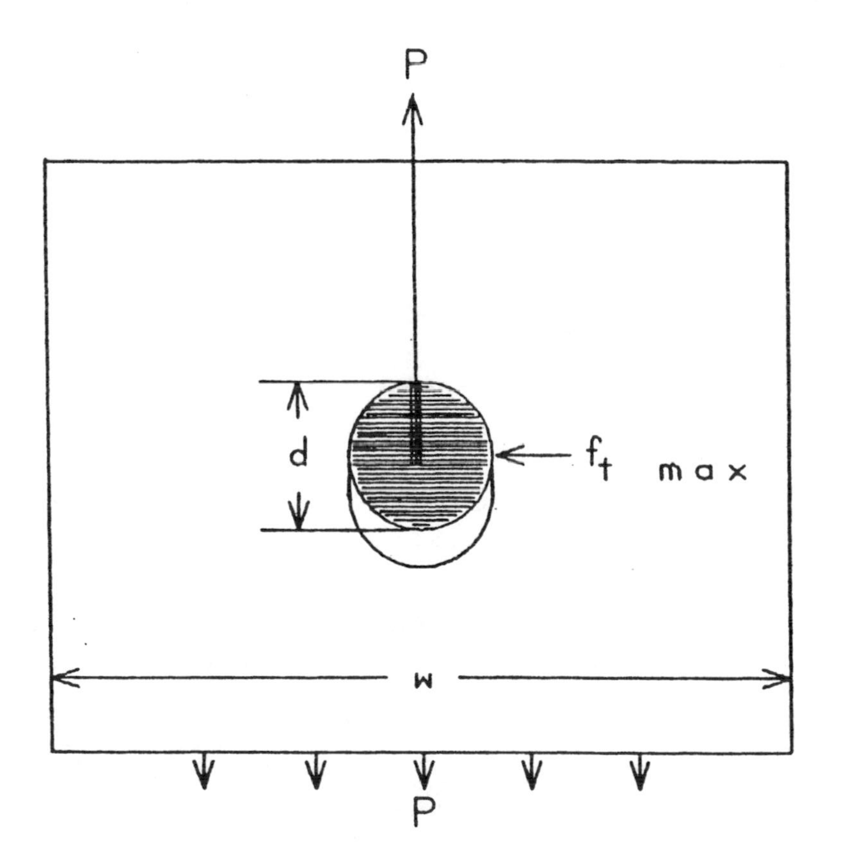


*Figure 12.7 Stress concentration in a pin-loaded hole.*

## *Stress Concentrations*

Sharp edge V-notches are avoided in design, but cut threads do provide this type of stress concentration.

Tool marks, caused by improper selection of pliers, pipe wrenches, water pump pliers, steel scribes, etc. can cause similar notches.

The stress concentration factor for a sharp edge notch is 6. Sometimes holes are not cleanly drilled and the drill hangs up as it exits the material. The resulting gouge in a drilled hole at the point of maximum tension stress results in a cumulative stress concentration and K = 6 x 3 = 18. These stress raisers are shown in Figure 12.8.

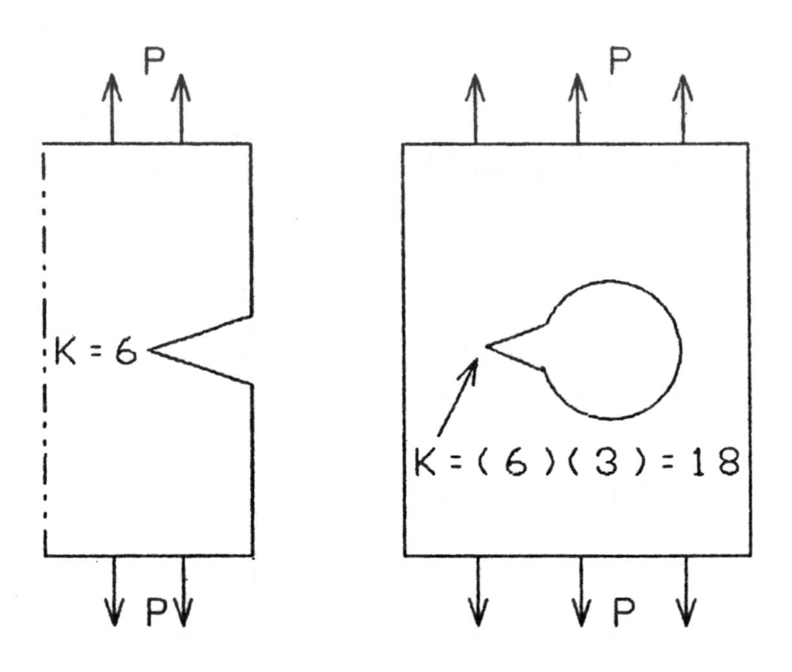


***Figure 12.8 Sharp V-notch stress concentration factors.***

As was shown in Figure 12.1, a sudden change in cross section produces a stress concentration. Fillets connecting the two section widths will reduce the stress concentration factors but not eliminate them completely.

The factors involved are: the radius of the fillet, r, the width of the larger section, W, and the width of the smaller section, w. Figure 12.9 shows the stress concentration factors for fillets on a flat bar under a tension load.

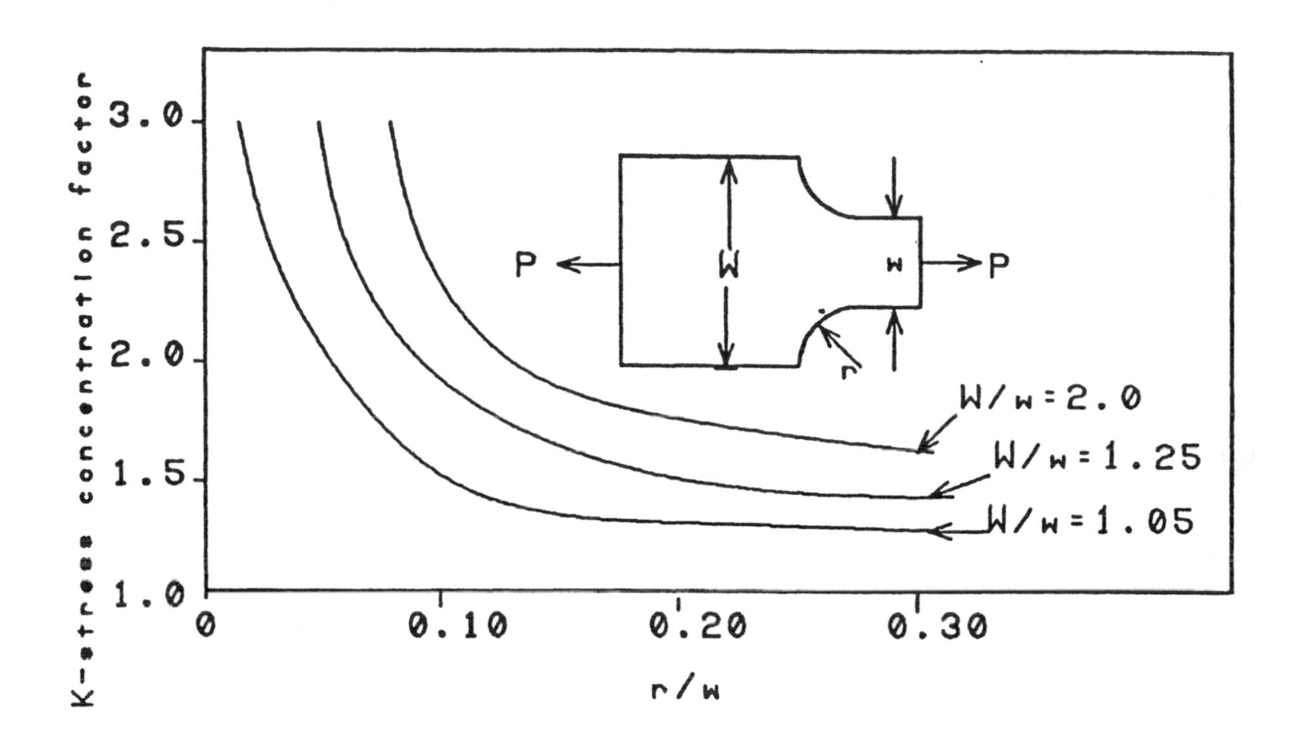


*Figure 12.9 Stress concentration factors for fillets on a flat bar under tension loading.*

There are many additional factors in an aircraft structure that influence local stress levels. Some of these are: residual stresses from fabrication and assembly, rubbing together of mating surfaces (fretting), welding, inclusions, etc. As users, rather than manufacturers, we must assure that maintenance procedures or malpractices and operational usage do not add new stress raisers to our equipment.

Stress concentrations lead to fatigue failures and fatigue failures can lead to disaster.

## SYMBOLS

$f_n$ = nominal stress, psi.

K = stress concentration factor

## EQUATIONS

12.1 $$K = \frac{\text{maximum actual stress}}{\text{nominal stress}}$$

12.2 $$K = 1 + 2\left(\frac{a}{b}\right)$$

## REVIEW PROBLEMS

1. Stress concentrations are more of a problem to static strength than to fatigue failures in brittle materials, because "K" remains constant in the elastic range and these materials are elastic until near the fracture stress.

(a) True
(b) False.

2. An elliptical hole in a wide plate with a width three times its height will produce a K value of:

(a) three.
(b) five.
(c) seven.
(d) one and two-thirds.

3. A hole with a sharp V-notch in it with the notch 90 degrees to the nominal stress of 1,000 psi. will develop an actual stress at the end of the crack of:

X 18

(a) 3,000 psi.
(b) 6,000 psi.
(c) 9,000 psi.
(d) 18,000 psi.

4. Ductile metals are primarily weakened by stress concentrations in:

(a) static strength.
(b) stiffness.
(c) fatigue.

5. Stress concentrations are more of a problem in ductile metals because:

(a) They greatly reduce the ultimate stress.
(b) They introduce cracks in the metal.
(c) They reduce the ductility of the metal.
(d) They increase the strain of the metal.

6. "Stop Drilling" is a time tested, effective way to stop fatigue cracks.

(a) True.
(b) False.

# CHAPTER THIRTEEN

# FATIGUE OF AIRCRAFT MATERIALS

Military specifications that cover the subject of repeated loads and fatigue are listed below.

| Mil Spec. | Subject |
|---|---|
| Mil-A-83444 | Airplane Damage Tolerance Requirements |
| Mil-A-008866B | Airplane Strength and Rigidity Reliability Requirements, Repeated Loads and Fatigue |
| Mil-A-8893 | Airplane Strength and Rigidity, Sonic Fatigue |

Similar civilian regulations are covered in sections 25.571 and 25.573 of Part 25 of Federal Aviation Regulations.

The design of "DAMAGE-TOLERANT" (fail-safe) structures has done much to alleviate fatigue failures in newer airplanes.

There are many older airplanes without the fail-safe features still flying and so fatigue failures persist. Mil-A-83444 and FAR 25.571 cover damage-tolerant structural design and will not be discussed in detail in this book.

Mil-A-8893 covers sonic fatigue. This subject is beyond the scope of this book and will not be discussed.

## SCOPE AND GENERAL REQUIREMENTS OF MIL-A-008866B

**SCOPE**- This specification identifies the durability design requirements applicable to the structure of airplanes. This specification applies to metallic and nonmetallic structures. The objective is to minimize the in-service maintenance costs and to obtain operational readiness through proper controls on material selection and processing, inspection, design detail, stress levels, and protection systems.

**GENERAL REQUIREMENTS**- The airframe shall be designed such that the economic life is greater than the design service life when subjected to the design service loads/environment spectra. The design objective is to minimize cracking or other structural or material degradation that could result in excessive maintenance problems or in functional problems.

# DEFINITION OF FATIGUE

*Fatigue* is the progressive localized structural damage. It occurs in a material subjected to repeated or fluctuating strains at stresses having a maximum value less than the ultimate tensile strength of the material.

## General Requirements for the Fatigue Process

There are three requirements for a fatigue crack to form and spread in metals:

1. There must be a **local** plastic stress.
2. There must be a tension stress.
3. There must be a cyclic (repeated or fluctuating) stress.

If we can eliminate any one of these three requirements, we can stop the fatigue process.

It should be noted that composite materials can fatigue under compression loads.

Military specification Mil-A-008866B states, "This specification applies to both metallic and nonmetallic structures." As metals are more susceptible to fatigue, we will concentrate the discussion on them.

# FATIGUE STAGES IN METALS

## Stage I: Crack Initiation

**Location:**

The first requirement gets the crack started. In Chapters Four and Twelve we discussed the movement of dislocations and resulting slip lines and slip planes. We said that these are cracks, thus plastic stress is required to move the dislocations and produce these cracks.

The emphasis on the word "local" in requirement 1 was to remind the reader that the entire structure does not need to be placed in the plastic range. Only one region of the structure or of a part must exceed the yield stress. This is why the importance of stress concentrations was emphasized in Chapter Twelve.

But, if dislocations are not considered as stress concentrations, it is possible that a fatigue crack can start, even without the presence of stress concentrations, if the nominal stress is in the plastic region. The crack, or multiple cracks, will start at the point where the local tensile stress is a maximum. The multiple cracks ultimately join to form a final crack.

**Orientation:**

The growth of a fatigue crack generally progresses by a step-wise shear process at approximately 45° to the axis of the tensile stress. At this point the crack is microscopic in size so this orientation will not be observable to the naked eye.

## Stage II: Crack Propagation (Fatigue zone)

The cyclic tension stresses spread the crack. If the part breaks in one application of the load, without the cyclic action of the stress, it is a static failure, not a fatigue failure.

**Orientation:**

Single cracks progress roughly perpendicular to the applied tensile stress. Multiple cracks parallel each other until final joining with other cracks. The line of joining is called a *herringbone pattern* and they point back in the direction of origin of the cracks (see Figure 13.1).

**Propagation Pattern:**

Fatigue cracks starting from a single point, with no other stress concentrations being present, propagate in a concave pattern from the point of origin (see Figure 13.1).

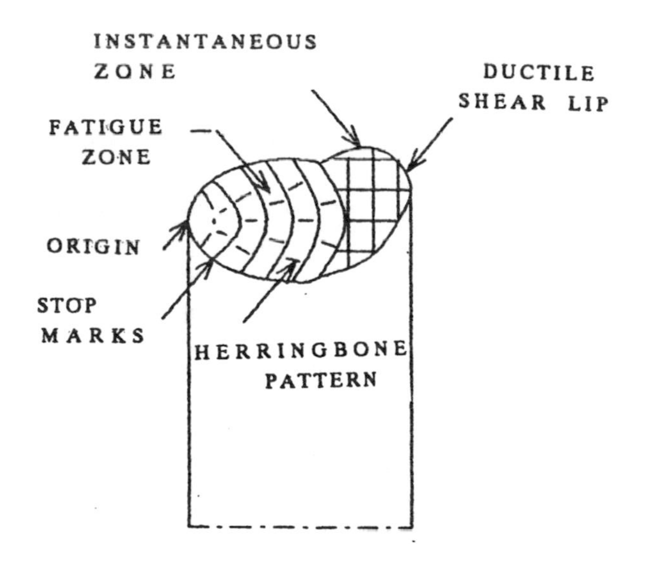


*Figure 13.1 Schematic of a typical fatigue failure.*

If a homogenous metal specimen is tested in fatigue in a testing machine, where the cycles of tension stress are uniformly applied at a constant rate, the fatigue crack will elongate at a fairly constant rate.

In actual practice this uniformity is missing. Most metals are alloys, thus the metal lacks homogeneity, particularly at the grain boundaries. The loads experienced in actual use are far from either being constant in value or in rate of application.

Thus a fatigue crack does not usually progress at a constant rate. It often stops entirely and starts up again if the load is increased. These movements are seen sometimes as concentric lines called *stop lines, beach marks, or clamshells.*

In this book we will call them "stop marks" as this describes their origin. They are shown in Figure 13.1. Stop marks are the most easily recognized characteristic of a fatigue failure. They may not always be seen with the naked eye, but may be seen under magnification.

It is rare that a cyclic tension load is applied without some compression resulting during the unloading process. This compression between the crack faces makes the fatigue crack smooth and polishes surface. In aluminum alloys this burnishing effect may obscure the stop marks.

## Stage III: Final Fracture (Instantaneous zone)

As the fatigue crack lengthens, there is a decrease in the amount of remaining metal to support the load. The stress in the remaining metal increases in value until the ultimate tensile strength of the metal is reached. At this point the specimen fails under the last single application of the load.

This is a static failure and closely resembles an ordinary static failure. Thus ductile materials have the final failure face near the 45° plane while the brittle metals will fail on the 90° plane to the applied load.

Figure 13.1 shows a ductile material rupture zone. In contrast to the smooth fatigue zone, the instantaneous zone will be rough in appearance.

## RECOGNITION FEATURES (More illustrations in the Appendix)

The origin of the crack, type of loading, direction of loading, and stress intensity can be determined from an investigation of the fatigue zone.

Schematic drawings of circular fatigue test specimens are shown in Figures 13.2, 13.3, and 13.4.
The sketches illustrate how the factors discussed in the paragraph above can be determined.

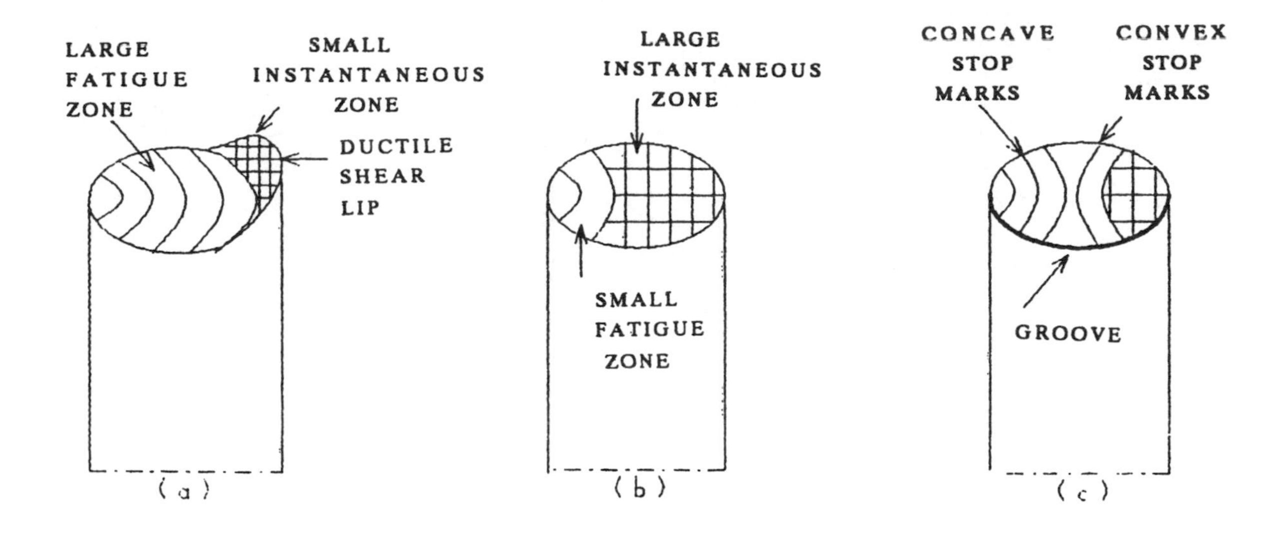


***Figure 13.2 Fatigue failures caused by cyclic tension loading or one way bending.***

Figure 13.2(a) shows a ductile metal having no apparent stress concentration. The ductility is shown by the shear type fracture surface. The stop marks and herringbone pattern point to the origin and show the direction of crack progression.

The single crack (or multiple cracks starting from near the same point) shows that tension or single directional bending was the direction of loading. The stress intensity is shown by the relative size of the fatigue zone and the instantaneous zone. If the stress intensity is low, many cycles are required for failure and the fatigue zone will be large.

Figure 13.2(b) shows a similar type failure except the fatigue zone is somewhat small, typical of a high stress level and fewer cycles required for failure. This is a brittle material as evidenced by the lack of shear lips and the 90° failure in the instantaneous zone.

Figure 13.2(c) shows that if an elongated stress concentration is present at the point of origin. This could be a groove cut around a circular specimen or a tool mark. The crack will progress along the stress concentration at a higher rate than through the interior of the metal and the concave pattern of the stop marks will be changed into a convex shape.

Figure 13.3 shows the failure pattern caused by two way bending. As each cyclic bending load is applied the opposite sides of the member are put in tension. A crack on either side can result from this action. If the cracks start opposite to each other the instantaneous zone will be between them.

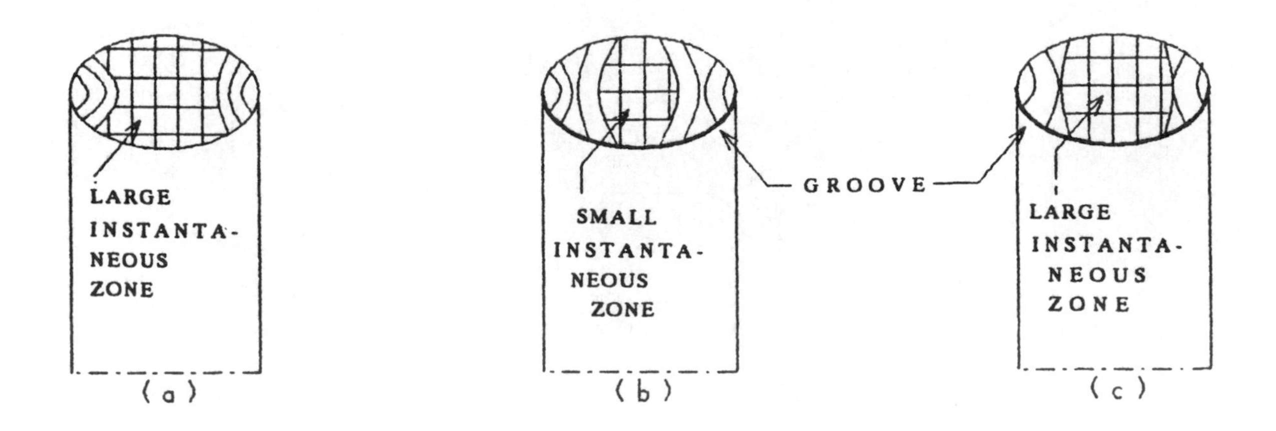


*Figure 13.3 Fatigue failures caused by two way bending.*

Figure 13.3(a) shows a brittle metal with no visible stress concentration. The stop marks starting from each crack origin maintain their concave pattern with respect to their origins until failure results somewhere near the center. This sample has a somewhat large instantaneous zone, again showing a high stress level with few cycles before failure.

Figures 13.3(b) and 13.3(c) show the presence of a stress concentration groove cut in the specimen. In both cases the concave stop mark pattern changes to convex before failure. The smaller instantaneous zone in 13.3(b) shows a lower stress level than the larger instantaneous zone in 13.3(c).

In Chapter Six we discussed how torsional loads produced, not only shear stress, but tension and compression stresses as well. The tension stress is of interest to us in the fatigue process. It is a maximum on the 45° plane, thus fatigue failures may occur on this plane due to torsional loading.

Figure 13.4(a) shows a schematic of a fatigue failure on a round drive shaft. The stress is a maximum at the fibers on the surface, therefore the crack will usually start along the surface before progressing inward. Without a stress concentration being present it will be difficult to pinpoint the exact origin of the crack. There will be herringbone type cracks pointing from the interior toward the surface and these help to identify the failure as a fatigue failure.

Presence of a stress concentration groove, as shown in Figure 13.4(b), alters the fatigue pattern drastically. The groove increases the local stress so greatly that the crack may follow the groove. The crack spreads inward only after going completely around the outer surface. If stop marks are present they form concentric rings moving inward on the 90° plane. The instantaneous zone will be located near the center of the shaft. Again the size of the instantaneous zone will show the stress level.

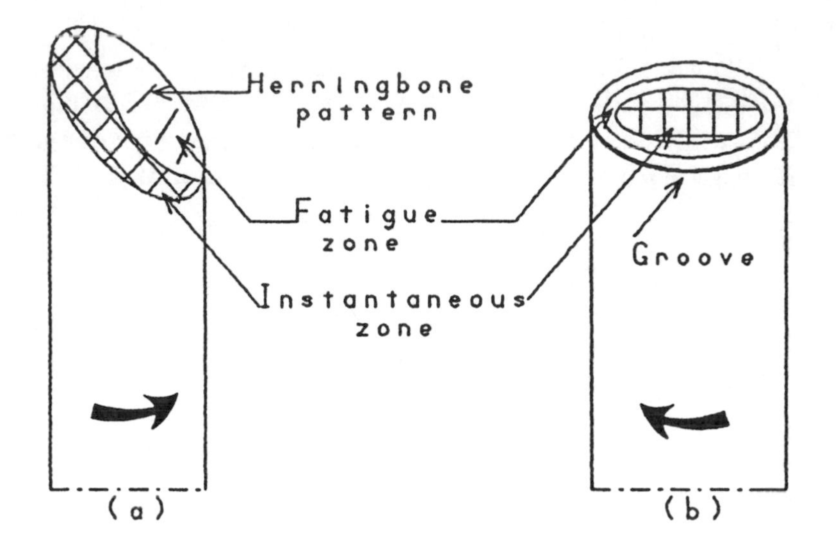


*Figure 13.4 Fatigue failures caused by torsional cyclic loads.*

## THE S-N DIAGRAM

A common mistake in the analysis of the fatigue process is to think of it as a function of time. The two most important factors are the tension stress level and the number of cycles. The stress vs. number of cycles (S-N) diagram, Figure 13.5, has been used as the primary source of fatigue failure prediction since fatigue testing was begun in 1858. Testing to failure often requires many cycles, so a logarithmic scale is most commonly used for the number of cycles. A linear scale is used for the stress amplitude.

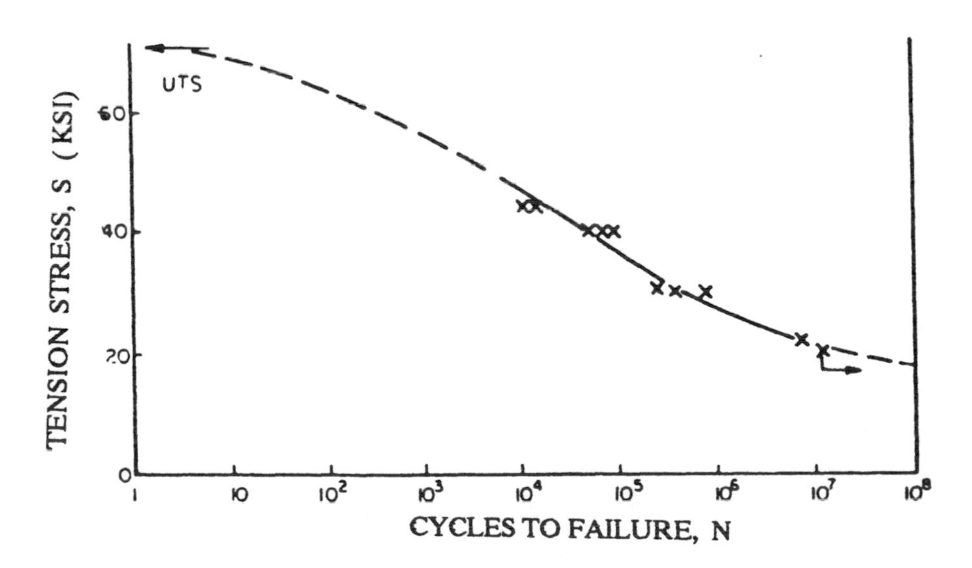


*Figure 13.5 Typical S-N diagram.*

Static stress tests produce failure at a fairly high degree of predictability. Fatigue tests, on the other hand, are less predictable and have a large "scatter" of results. A typical S-N curve for an aluminum alloy specimen is shown in Figure 13.6. The dashed lines represent the minimum and maximum number of cycles for failure at several stress levels. The solid line represents the mean curve where the survival probability is 50 percent.

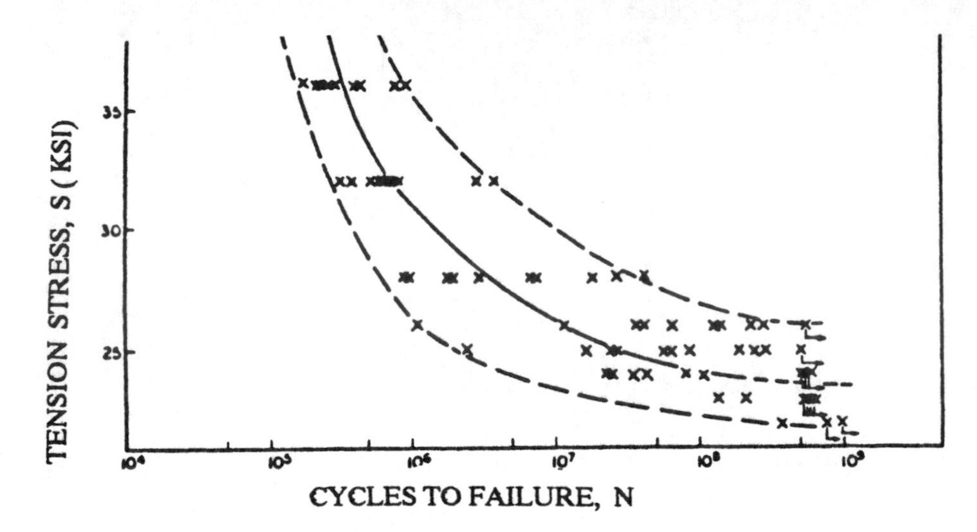


*Figure 13.6 Scatter of S-N values.*

The use of statistical reasoning permits us to formulate probability of survival (P-S-N) curves. Typical P-S-N curves are shown in Figure 13.7.

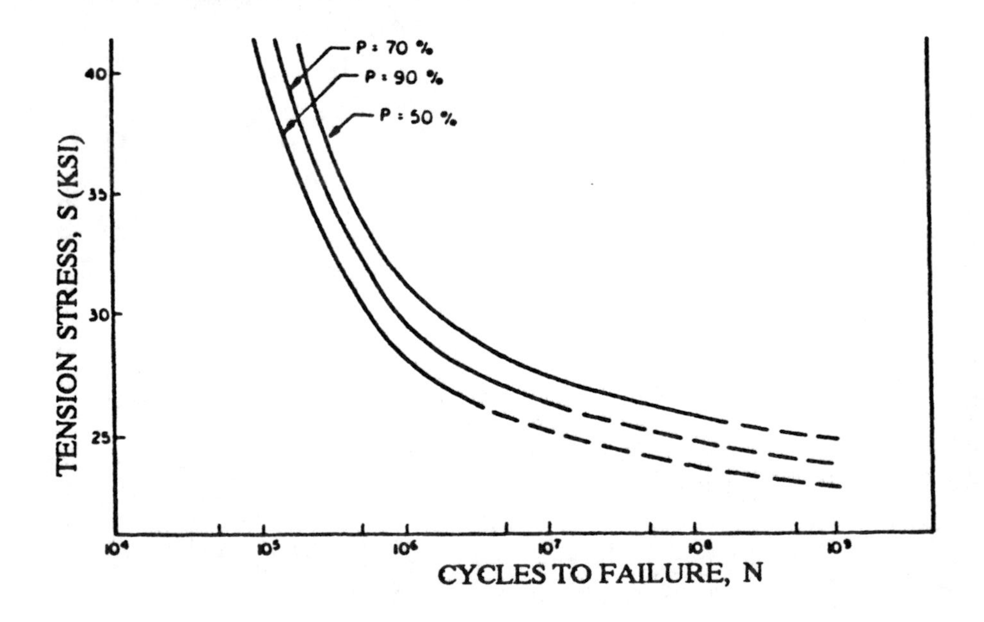


*Figure 13.7 P-S-N curves.*

Stress concentrations lower the curves. Affect of notches on the S-N diagram is shown in Figure 13.8

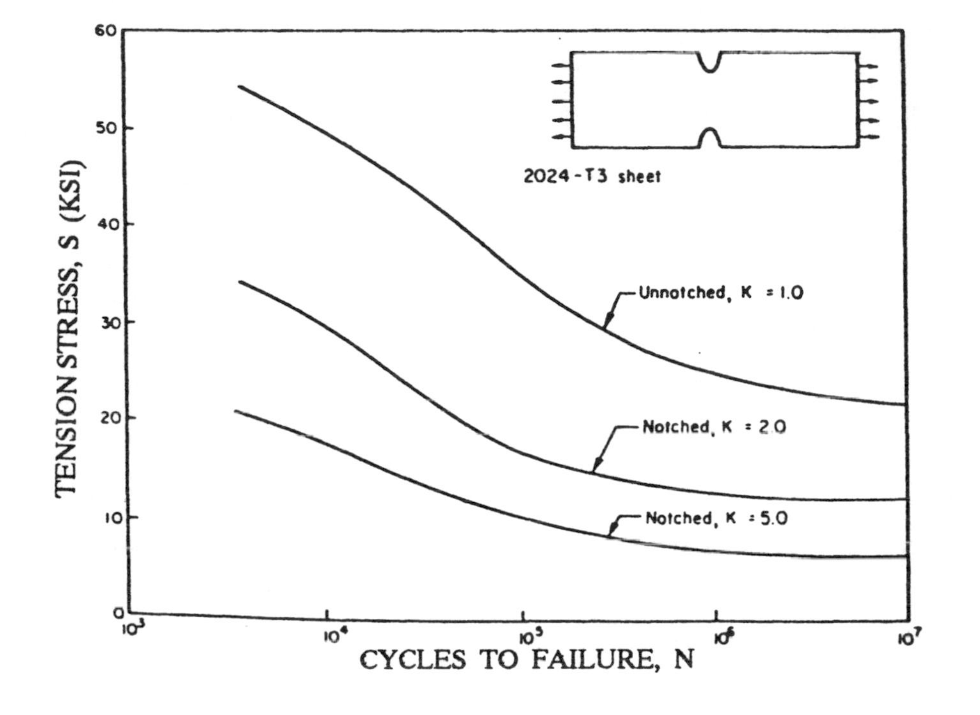


***Figure 13.8 Affect of notches on fatigue strength.***

## ENDURANCE LIMIT

The S-N curves flatten out at lower stress limits. This shows a limit of stress below which fatigue failure will not occur. This is called the *endurance limit* or fatigue limit. Steel, titanium and their alloys exhibit a definite endurance limit. Aluminum, magnesium and other ductile materials do not exhibit this property. Theoretically, they will fail under any load, if enough cycles are applied. From a practical standpoint it is usual to define an endurance limit for these materials at $10^7$ or $10^8$ cycles.

While the S-N diagram is useful in predicting fatigue life of individual aircraft parts, an entire aircraft is much more complex. Fatigue testing of an entire structure, under conditions simulating actual flight conditions, is a complicated and time consuming process. This will be discussed later.

## CUMULATIVE DAMAGE THEORIES

Many theories have been advanced concerned with the fatigue life of materials subjected to fluctuating loads. Most of these theories are based upon laboratory data in which the stresses are applied in a sine wave pattern. A detailed analysis of the theories is beyond the scope of this book, but a simple explanation of Miner's theory seems warranted.

Consider that the two-level sinusoidal stress history shown in Figure 13.9 with maximum stress levels $S_1$ and $S_2$ are applied to a material alternately in groups of $n_1$ and $n_2$ cycles respectively.

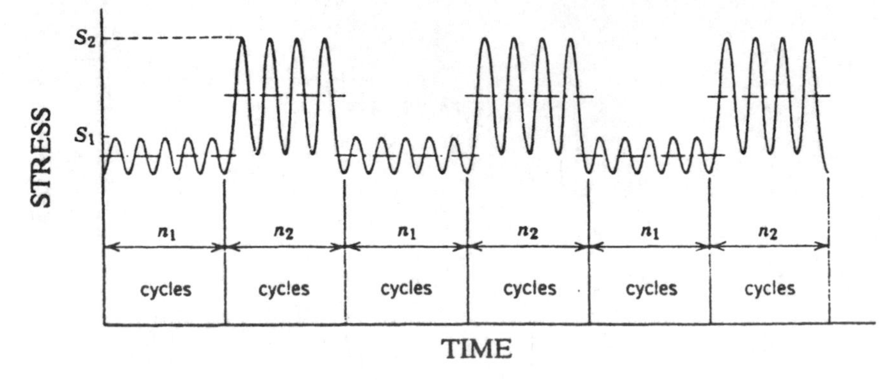


*Figure 13.9 Two level sinusoidal stress history.*

The number of cycles to failure at stress levels one ($S_1$) and two ($S_2$) are called $N_1$ and $N_2$ respectively. The amount of damage, $D_1$, done by stress level $S_1$ applied $n_1$ cycles is:

$$D_1 = \frac{n_1}{N_1}$$

Similarly the damage, $D_2$, done by stress level $S_2$ applied $n_2$ cycles is

$$D_2 = \frac{n_2}{N_2}$$

Total damage, $D_T = D_1 + D_2$.

The theory is that when $D_T = 1.0$, failure will occur.

## FULL-SCALE AIRCRAFT FATIGUE TESTING

Preventing fatigue failures in aircraft structures can be enhanced by design, testing of small components, and in care during production. There are complex interactions between the various assembled components that can only be determined by full-scale fatigue testing. Mil-A-008866B and Mil-A-8871A list the requirements for such testing.

Limitations of tests include: (1) the complete assembly can have only a single test, (2) the test is usually an accelerated test and, (3) the uncertainty of the test program.
An example of this last point is that mission requirements often change after the test is completed.

The objectives of full aircraft fatigue testing are: (1) identifying critical fatigue areas, (2) indicating modes of potential failure, (3) estimating service lifetime before fatigue failure, (4) estimating fail-safe characteristics.

Mission profiles such as shown in Figure 13.10 are used to provide a basis for defining the fatigue test stress levels.

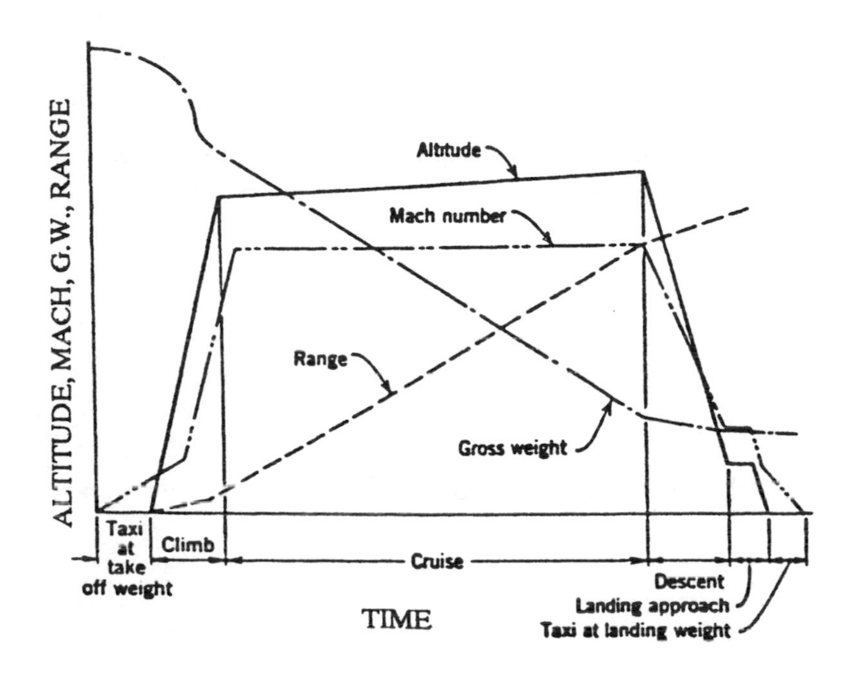


*Figure 13.10 Flight profile for cargo airplane.*

From the flight profile the load spectrum can be predicted. The complete spectrum for a certain mission is known as the ground-air-ground (GAG) cycle. Typical GAG cycles of loads are shown in Figure 13.11. This spectrum is entered in a computer and then becomes the test cycle.

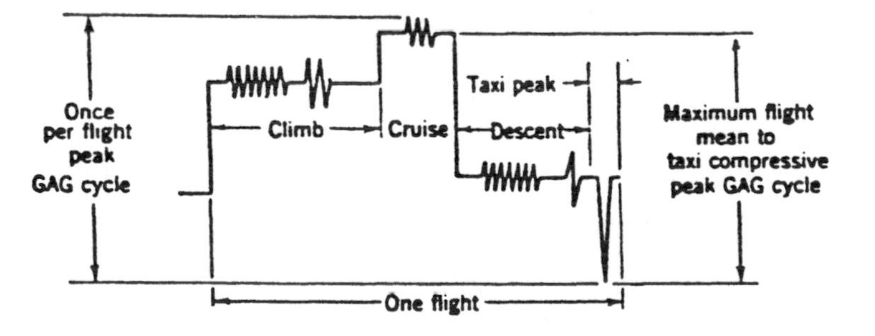


*Figure 13.11 Flight-by-flight low-high ordered loading sequence.*

## AVOIDANCE OF FATIGUE FAILURES

The principal approaches to the problem of preventing fatigue failures are discussed below.

### 1. Choice of Materials with Inherent High Fatigue Resistance

As we discussed in Chapter Eight, the toughness of a metal depends upon its tensile strength and its ductility. These properties can be altered by either changing the alloy or by heat treatment. But, increasing one of these properties also affects the other.

Increasing the strength of a metal results in decreasing its ductility and vice versa. For high imposed strain (1 to 2 percent or more) the ductility is most important. High ductility will prolong the life. Below this strain range an increase in tensile strength will prolong the life.

The introduction of sharp stress raisers complicates the picture. It has been found that when sharp notches are present that the tougher material will have a longer fatigue life than the less tough material.

The definition of toughness that we used in Chapter Eight was, "the ability to absorb energy before failure" does not apply to fatigue failure. A correct definition would be, "the ability to carry a cyclic tension load in the presence of a stress concentration." Proper selection and heat treatment of materials have a strong influence on the fatigue life of a structure.

Laminated composite materials have good resistance to fatigue damage, if the lamina fibers are oriented in the proper direction. The fibers are good "crack stoppers" if they are oriented at 90° to the crack direction as shown in Figure 13.12.

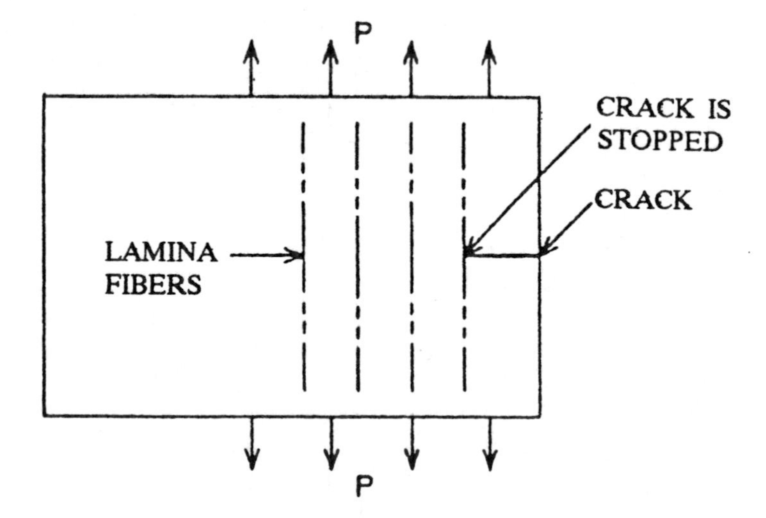


*Figure 13.12 Laminated composites stop fatigue cracks.*

## 2. Avoidance of Stress Concentrations

The subject of stress concentrations was discussed in some detail in Chapter Twelve. Maintenance personnel must be educated to the dangers of introducing stress concentrations such as tool marks, scratches, dents, and similar maintenance malpractices. Pilots also must be aware of these hazards and promptly report observed stress concentrations caused by personnel or foreign object damage (FOD).

## 3. Surface Protection

Nearly all fatigue cracks start at the surface of a body. The absence of restraining atoms outside of the surface atoms changes the cohesive forces, removes constraints and generally makes the surface "weaker" than the interior. The outer surface is also subject to stress concentrations, tool marks and rough handling. Thus, protection of the surface will improve the fatigue resistance.

## 4. Favorable Residual Stress

The most commonly used method of improving fatigue life is the introduction of residual compressive stress. Since tension stress is required to spread fatigue cracks in metals, placing the metal in compression will reduce or eliminate entirely the tension stress. Several methods of introducing residual compressive stress are discussed here.

a. *Shot peening.* The surface to be protected is bombarded by lead shot or glass beads. Each shot puts a small compressive dent in the surface of the material. The dents will overlap and the entire surface will be in compression.

b. *Surface hardening.* Atoms of the hardening material are introduced into the surface metal. These force the parent atoms apart and the surface tries to expand. Expansion can take place in the outward direction but is restrained by the sub-surface metal from expanding in the direction of the surface. This puts the surface into a state of compression.

In addition, the hardened surface resists scratches and tool marks. This is a bonus effect in fatigue protection.

c. *Rolled threads.* The sharp roots of machine cut threads introduce a stress concentration. Rolled threads are formed by forcing the metal into the shape of the threads, instead of cutting the metal away.

This creates fatigue resistance in two ways:

First, the stress concentration factor is sharply reduced in value.
Second, the roots are forced into compression by the rolling process.

Figure 13.13(a) shows cut threads and 13.13(b) shows the improvement in fatigue life of rolled threads over cut threads.

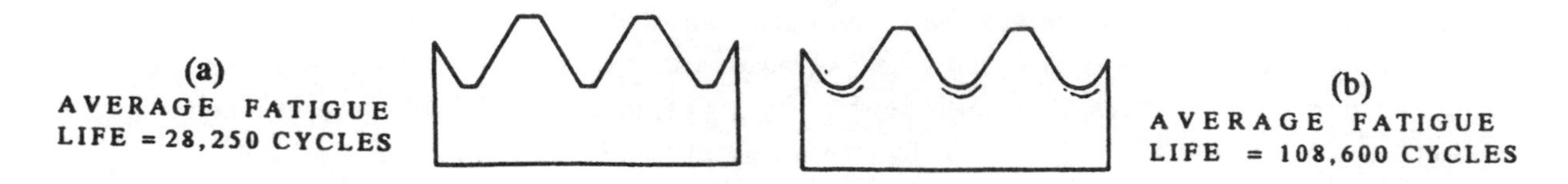


*Figure 13.13 (a) cut threads; (b) rolled threads.*

*d. Hole protection.* If the sides of a hole are put in compression the local tension stress caused by the stress concentration factor (K = 3) will be sharply reduced. There are several ways of doing this. One way is by using a rivet in the hole. Upon driving, the rivet expands and compresses the sides of the hole. If bolts are used to fasten the joint no such compression is produced. Holes can be expanded by drilling the hole slightly smaller than the bolt and then enlarged by forcing a mandrel through the hole. More sophisticated methods have been invented by commercial firms. One of these is called Taper-Loc ®. A tapered bolt is inserted in a tapered hole. A combination washer and nut is tightened that pulls the bolt into the hole, thus inducing compressive stresses in the sides of the hole. This process is shown in Figure 13.14.

**Taper-Lok®** Installation

Step 1. Tightly clamp pieces together.

Step 2. Drill pilot hole.

Step 3. Ream hole with tapered reamer.

Step 4. Drop fastener into hole.

Step 5. Press fastener head down with finger pressure.
Check head protrusion with gage.

Step 6. Start washer-nut on by hand.

Step 7. Tighten nut by conventional wrenching methods.

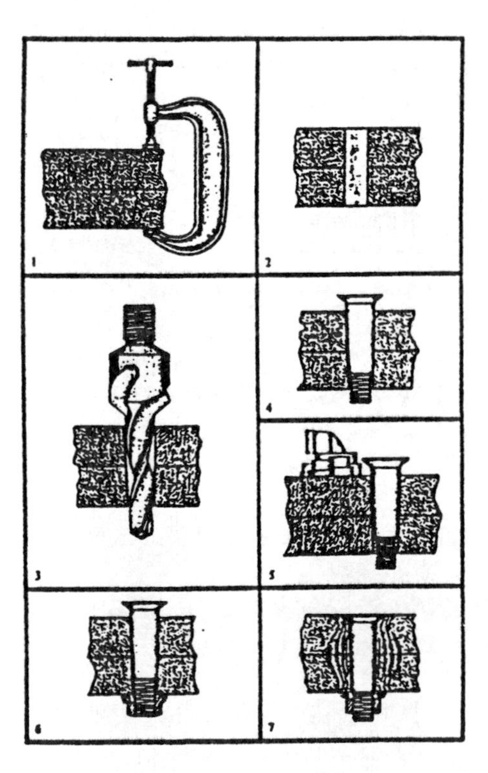

*Figure 13.14 Taper-Loc® installation.*

The same principle is used in the CX PROCESS® Cold Expanded Hole developed by Fatigue Technology, Inc. In this process a non-tapered bolt hole is drilled and a split bushing inserted in the hole. A special mandrel forces the bushing open, pressing against the sides of the hole and introducing compressive stress. Figure 13.15 is a schematic drawing of the sequence of the process.

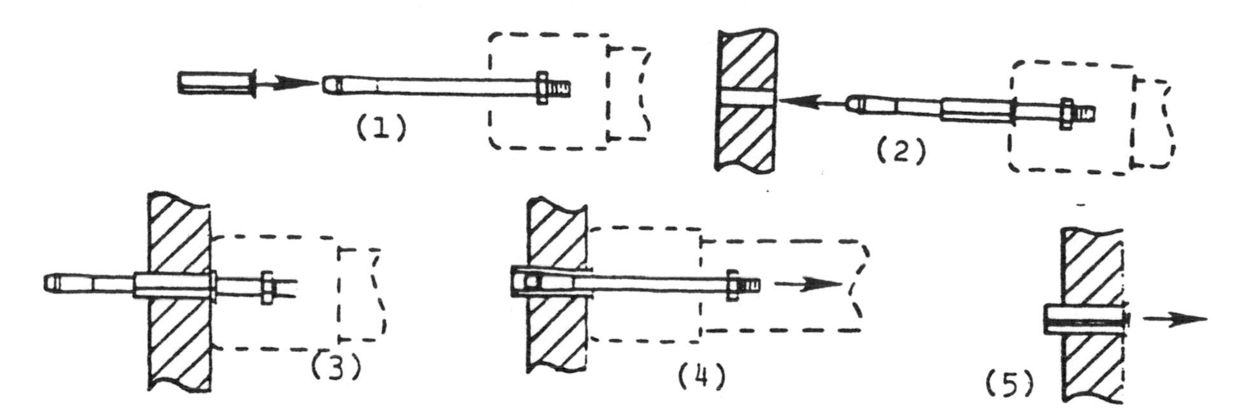


*Figure 13.15 CX PROCESS® installation.*

Figure 13.16(a) shows the stress level at the side of the hole after the bushing has been expanded but before any load has been applied to the metal plate. Figure 13.16(b) shows the stress at the side of the hole after a 20,000 psi. tension load has been applied.

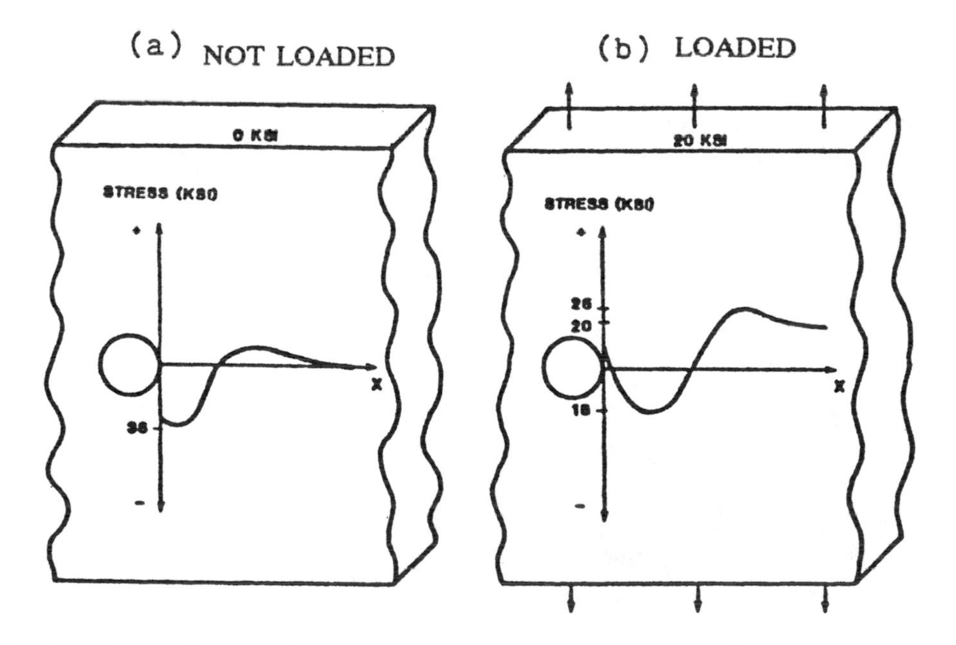


*Figure 13.16 Stress distribution of CX PROCESS ® cold expanded hole.*

## 5. Reduction of Cyclic Stress

*Preloaded fasteners.* These are more commonly called *Torqued bolts.* Bolted joints are fastened in two ways:

The lap joint shown in Figure 13.17(a).
The tension butt joint shown in Figure 13.17(b).

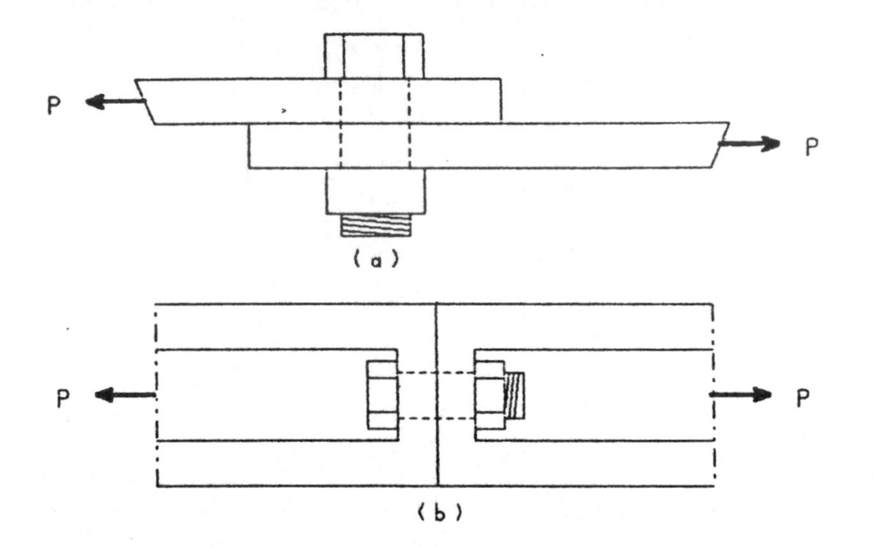


*Figure 13.17 Types of bolted joints.*

Torque may be defined as the moment caused by a force that causes rotation. The torquing of a lap joint bolt produces friction between the mating surfaces. This friction will carry part of the applied load so the shear load on the bolt is reduced.

Thus the hole itself will be subject to reduced load and the stress concentration factor of the hole will be reduced. This was shown in Figure 12.7.

The tension butt joint is extremely critical. It is avoided in design whenever possible. Here all tension loads on the mating surfaces must be carried by tension stresses in the bolt. These loads are nearly always cyclic loads, sets up the classical requirements for fatigue.

Consider a bolt that is holding a tension butt joint and the nut is screwed on hand tight (with no torque being applied). All fluctuations in load on this joint result in cyclic stressing of the bolt. This is one requirement for fatigue.

Let's see how cyclic stress can be reduced.

With no load on the joint, torque (preload) the bolt to a tension stress of below the yield stress of the bolt.

The bolt is now in tension and the mating surfaces are in compression. In order for applied tension loads to increase the tension stress in the bolt, the bolt head and the nut must be moved farther apart.

This can only be done by decreasing the compression stress in the mating surfaces, thus allowing them to expand. Mating surfaces have large contact areas compared with the bolt cross sectional area.

Therefore, with an applied load, the mating compressive stress ($f = P/A$) will be smaller than the bolt tension stress by the ratio of the areas. Fluctuations in load result in smaller stress changes in the compressed faces.

If both parts are made of the same material the strain also will be much less for the faces and the cyclic loads produce only small movements in the length of the bolt and stress variations in the bolt are greatly reduced.

## 6. Fail-Safe Design

While this book is not intended to discuss the design aspects of material factors, a short summary of the fail-safe philosophy seems warranted. The simple rules of the philosophy are:

1. The structure must have adequate life, either as a crack-free period, or one during which the growth rate of cracks is sufficiently low to escape detection by inspection procedures.

2. The structure must be able to carry a designed load with a given amount of damage.

3. Visual inspection of critical areas must be possible.

4. Repair of damaged structure must be possible either by replacement or by use of patching doublers.

## 7. Inspections

Often the fatigue crack cannot be seen until late in the part's fatigue life. For a smooth specimen this may mean as much as 90 percent of the life has been used.

It should be obvious that it is necessary to have inspection periods scheduled to ensure that the number of cycles between inspections does not exceed the number of cycles between detection and failure.

Inspections must be thorough and non-destructive inspection procedures (to be covered in Chapter Fifteen) be used on known or suspected fatigue crack locations.

## REVIEW PROBLEMS

1. There are three factors that must exist if a fatigue failure is to occur in metals. These are:
   (a) tension stress, cyclic stress, average overstress.
   (b) tension stress, cyclic stress, local plastic stress.
   (c) stress concentration, cyclic stress, tension stress.

2. If a fatigue crack starts without an external stress concentration being present, the most common reason is:
   (a) impurities in the metal cause cracks.
   (b) dislocations in the metal move and cause cracks.
   (c) internal stresses from heat treating cause cracks.

3. True-False. Fatigue microcracks widen to become macrocracks and when these can be detected there is still about half of the service life remaining.

4. True-False. It is necessary to have external stress concentrations present for fatigue cracks to start.

5. The stress below which a fatigue crack will never start is called:
   (a) yield stress
   (b) proportional limit
   (c) ultimate strength
   (d) endurance limit

6. Calculate the fatigue damage according to Miner's theory for the below stresses and S-N diagram.

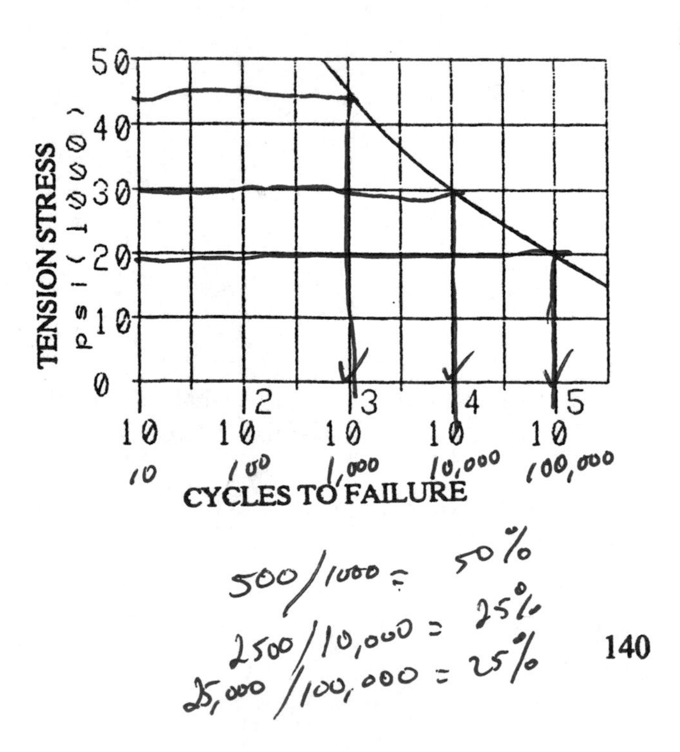


| Stress | Cycles | Damage |
|---|---|---|
| 45,000 | 500 | 50% |
| 30,000 | 2500 | 25% |
| 20,000 | 25,000 | 25% |

Total Damage 100%

Amount of service life used up?

(a) 100%
(b) 75%
(c) 10%
(d) 1%

# CHAPTER FOURTEEN

# CORROSION

Aircraft metals are light strong alloys that are highly susceptible to corrosion. There is no known method of eliminating all corrosion problems, but there are methods of protecting the metals and thus controlling the corrosion process. Typical protecting measures are: sealing, painting, and chemical treatment of the metallic parts.

Though the designer and manufacturer take extensive corrosion control measures, once an airplane enters service it is exposed to corrosive attack and some damage will occur. The corrosion must be identified promptly and corrective action taken. Effective corrosion control consists of: keeping the aircraft clean, maintaining the protective coatings, and conducting frequent inspections to detect and arrest corrosion as soon as possible.

## DEFINITION OF CORROSION

Corrosion in metals can be defined as, "The disintegration of a metal that results from the interaction of metallic surfaces with one or more substances in the environment."

This interaction is affected by such factors as temperature, stress and fatigue loading. The result of this interaction is the transformation of the metal into chemical compounds. These, when dry, are usually brittle, scaly, or powdery, and have little mechanical strength.

Moisture is the source of most aircraft corrosion problems. Most authorities consider moisture to be a corrosive agent by itself. Corrosion control should begin with keeping metallic surfaces clean and dry.

## GENERAL TYPES OF CORROSION ARE:

(1) Chemical corrosion

(2) Electrochemical corrosion.

The above classifications may be criticized on the basis that there is electrochemical reaction present in practically all corrosion.

But, in chemical corrosion, the distances between the anodic (positively charged) areas and cathodic (negatively charged) areas on the metallic surface are so small that there is no perceptible current flow.

## CHEMICAL CORROSION

### Direct Chemical Attack

Direct chemical attack proceeds at a nearly even rate over the surface. A common example of direct chemical attack is the reaction of iron with moist atmosphere that produces rust. Another example is a corrosive agent such as sulfuric acid pickling solution used to clean steel surfaces. The surface is dissolved uniformly without the formation of protective layers and the attack continues at an almost constant rate.

A less severe example is that of aluminum exposed to air. Aluminum oxide forms on the surface and, after it thickens, forms a barrier to the air and the corrosion process stops. The aluminum oxide is unsightly and is often removed from clad aircraft skin by polishing. As the pure aluminum cladding is very thin, continued polishing may remove the aluminum and expose the aluminum alloy base metal to the air. The exposed aluminum alloy is subject to electrochemical corrosion as intergranular corrosion, which will be discussed later. Figure 14.1 shows a photomicrograph of a clad aluminum alloy sheet. The thinness of the cladding should be noted.

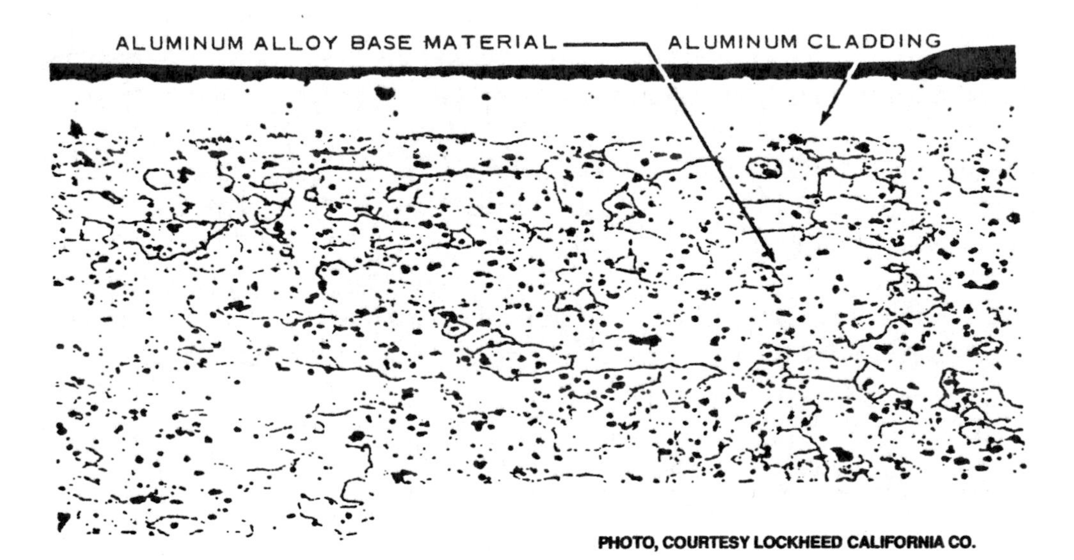


***Figure 14.1 Aluminum clad aluminum alloy sheet.***

### High-Temperature Oxidation

This type of corrosion involves the reaction of metals with oxygen at high temperatures, usually in the absence of moisture. Engine and auxiliary power unit exhaust areas are affected by a discoloration or general dulling of the surface.

## ELECTROCHEMICAL CORROSION

Metals vary in their resistance to corrosion. A Galvanic Series of metals is shown in Table 14.1.
The metals are listed in the order of their tendency to corrode. The more anodic a metal is, the greater its galvanic corrosion will be when it is in contact with a less anodic (more cathodic) metal.

**TABLE 14.1 Galvanic Series of Metals**

| Anodic. Least Noble. Corroded End |
|---|
| Magnesium |
| Magnesium alloys |
| Zinc |
| Aluminum alloys (low strength) |
| Cadmium |
| Steel or iron |
| Cast iron |
| Stainless steels (active) |
| Lead |
| Tin |
| Nickel (active) |
| Brasses |
| Copper |
| Bronzes |
| Copper-nickel alloys |
| Nickel (passive) |
| Silver |
| Titanium |
| Graphite |
| Gold |
| Platinum |
| Cathodic. Most Noble. Protected End |

There are four conditions that must exist before electrochemical corrosion can occur.
(1) There must be something to corrode (the anodic metal);
(2) There must be a cause for corrosion (the cathodic metal);
(3) There must be a continuous liquid path (the electrolyte, salt water or other contaminants); and
(4) There must be a conductor to carry the flow of electrons from the anode to the cathode. This conductor is created when metal touches metal as around rivets, bolts, and welds.
The four conditions that are necessary before electrochemical corrosion can proceed are shown in Figure 14.2.
The elimination of any of the four conditions will automatically stop corrosion.

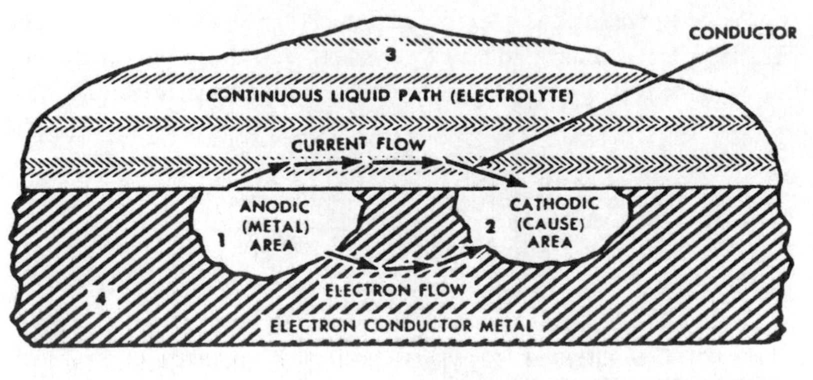


PHOTO, COURTESY NORTHROP CORP.

*Figure 14.2 Conditions required for electrochemical corrosion.*

A method of preventing corrosion by preventing the electrolyte from connecting the cathode and anode is by applying an organic film to the surface. This is shown in Figure 14.3. Anodizing or cladding the surface will have a similar effect.

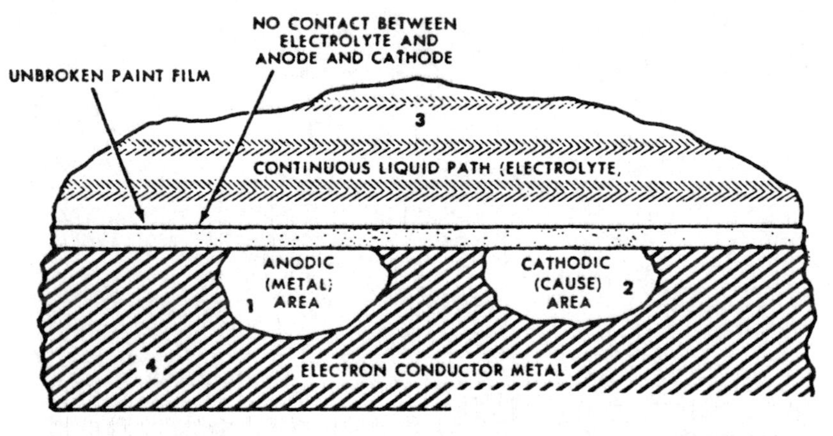


PHOTO, COURTESY NORTHROP CORP.

*Figure 14.3 Organic film prevents corrosion.*

## FORMS OF CORROSION (More illustrations in the Appendix)

### Uniform Attack

This is any form of corrosion where the whole surface of the metal corrodes to the same degree. While this form of corrosion is unsightly, it is also easily detectable. Uniform attack is usually due to chemical attack, although electrochemical attack is not uncommon.

## Localized Corrosion

Corrosive attack is often localized in well defined areas is not spread uniformly over an entire surface. All forms of localized corrosion are of an electrochemical nature and fall into two distinct categories: *pitting* or *selective attack.*

## Pitting

This type of corrosion is a case of localized attack confined to very small areas of the metal surface. The remainder of the surface is not affected.

Usually the pits are randomly located over the surface, although some preferential attack may occur at the grain boundaries of metal. A typical pit has a sharp, well-defined edge with walls that run almost perpendicular to the metal surface. Some pits undercut the metal surface and branch out like roots of a tree.

All pits have one thing in common. They can penetrate deeply into a piece of metal and cause damage that is completely out of proportion to the amount of metal consumed.

Pitting results from the chemical action of moisture, acid, alkali, or saline solutions on the metal, after the protective coating has been removed or penetrated.

Pitting of clad aluminum alloy is shown in Figure 14.4. The cladding has not been completely penetrated at this point.

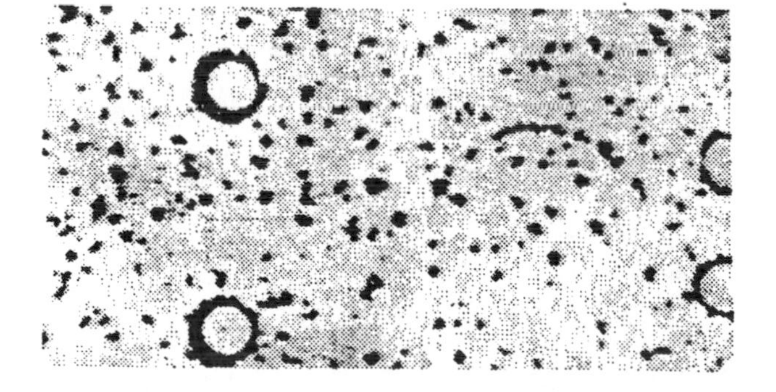

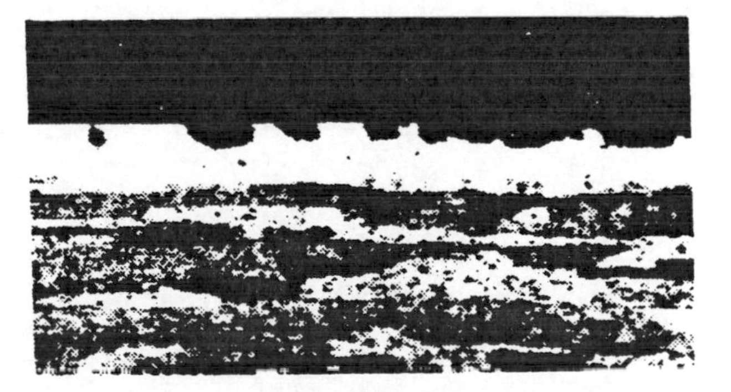

PHOTO, COURTESY LOCKHEED CALIFORNIA CO.

*Figure 14.4 Macrograph and micrograph of pitting of clad aluminum alloy panel.*

## Selective Attack

Some forms of corrosion seem to discriminate against one particular phase or constituent of an alloy while other phases are apparently untouched. This type of corrosion is called *selective attack*.

## Intergranular Corrosion

*Intergranular corrosion* attacks one particular phase or constituent of an alloy while ignoring other phases. The attack generally begins with pitting and progresses inward until it reaches the boundaries between grains.

Although we have stated that two dissimilar metals are required for electrochemical corrosion, an alloy has different chemical composition of the metal grains and the grain boundaries. These are, in effect, dissimilar metals. The grain boundaries are anodic with respect to the grains and the boundaries are corroded away.

The strength of the alloy depends largely on the chemical binding effect of the grain boundaries. Once the grain boundaries are destroyed the grains are still mechanically locked together, but the chemical bonds are broken. Like pitting, the damage of this form of attack causes a loss of strength and ductility out of proportion to the amount of metal destroyed. If an electrolytic medium, such as salt-air moisture, is present a metallic alloy can literally destroy itself. The eroding of the aluminum cladding of an aluminum alloy by constant polishing (mentioned earlier) can lead to this type of corrosion.

Figure 14.5 shows intergranular corrosion of 2024-T4 aluminum alloy.

PHOTO, COURTESY LOCKHEED CALIFORNIA CO.

***Figure 14.5 Micrograph of intergranular corrosion.***

## Exfoliation

*Exfoliation* is a form of intergranular corrosion in which the corrosion progresses parallel to the surface of the metal. The corroded metal peels out in layers as shown in Figure 14.6. This type of corrosion often exists in extrusions. The extruding process elongates the grains of the metal parallel to the surface. The products of intergranular corrosion take up more volume and forces the surface layers outward. Raised surfaces near bolt or rivet holes are the clue to this form of corrosion.

PHOTO, COURTESY LOCKHEED CALIFORNIA CO.

*Figure 14.6 Macrograph and micrograph of exfoliation corrosion of extruded 2024-T4 Al alloy channel.*

# CORROSION AND MECHANICAL FACTORS

Mechanical factors such as stress, fatigue loading or erosion can aggravate corrosive attack. These factors can either be applied to the part or can be residual, as from poor heat treatment techniques. They can be static or cyclic stress forces. In any case they cause the corrosion to progress at an accelerated rate. Examples of these interrelationships are discussed below.

## Stress-Corrosion Cracking

The combined action of statically applied tension loads or residual tension stresses and corrosive attack can cause stress-corrosion cracking. There are several facts that characterize this type of corrosion:

First, the damage is caused by the combined action of tension stress and corrosion applied simultaneously.

Second, the actual event occurs in two stages:

(a) the period of crack initiation. Generally, the crack initiation results from a physical breakdown of protective surface films and the subsequent corrosive attack on the part.

(b) the period of crack propagation. Crack propagation often involves an electrochemical attack on the surfaces of the crack, particularly at the apex of the crack where the stress is highest.

Third, the rate of corrosive attack on the sides of the crack is low in comparison to the rate of attack at the apex of the advancing crack.

Fourth, the crack may be either intergranular or transgranular.
Figure 14.7 shows stress-corrosion cracking of a 7075-T6 aluminum alloy part.

PHOTO, COURTESY LOCKHEED CALIFORNIA CO.

*Figure 14.7 Macrograph and micrograph of stress-corrosion cracking of 7075-T6 Al alloy part.*

## Corrosion Fatigue

The simultaneous action of corrosion and metal fatigue is called *corrosion fatigue*. When conditions conducive to fatigue failure and corrosion act simultaneously, the predicted fatigue life of a structure or part can be significantly shortened. Pits, as discussed earlier, are stress raisers and may easily be the beginning of fatigue cracks. Figure 14.8 shows corrosion fatigue cracks starting from corrosion pits in a bolt hole.

PHOTO, COURTESY BOEING WICHITA

*Figure 14.8 Corrosion fatigue cracks.*

## Fretting Corrosion

Damage can occur at the interface of two highly loaded surfaces that are in compression, if they are subject to vibration. The vibration can cause the surfaces to rub together, resulting in abrasive wear that removes the protective oxide films and part of the virgin metal. This type of wear is a form of erosion known as *fretting*.

Although the fit between the two surfaces may be extremely tight, it is rarely tight enough to prevent oxygen or another corrosive agent from entering and attacking the unprotected surfaces. This will cause additional destruction of the metal. This combined action is called *fretting corrosion*.

If the amplitude of motion is not very large, the debris (oxides and metal particles) formed by fretting corrosion has little opportunity to escape from the interfacial area.

The abrasive action of the debris will form more abrasive material and accelerate the mechanical wear until a path is finally formed to permit escape of the debris. Usually, before this path can be formed, the surfaces have suffered extreme wear and the parts must be replaced.

Many basic principles of fretting corrosion are not fully understood, and knowledge is limited mostly to the field of ferrous materials.

But, it is believed that the characteristic features of the fretting corrosion problems encountered with iron and its alloys are common to other materials. Currently, fretting corrosion is considered to be predominantly mechanical wear combined with some corrosive attack.

# CORROSION DETECTION

The stealthy nature of corrosive attack makes the use of aggressive corrosion detection measures essential.

Visual inspection, the primary approach to corrosion detection, is indispensable, but when visual inspection is not feasible other techniques can be applied.

Some success has been achieved by the use of liquid penetrant, magnetic particle, X-ray, and ultrasonic detection methods. But most of these sophisticated methods are aimed toward the detection of physical flaws within metal objects.

So, the success of these methods has usually been limited to the detection of advanced corrosive attack such as stress-corrosion cracks, corrosion fatigue cracks, and exfoliation.

Consequently, we still must rely upon visual inspection to find corrosive attack during its incipient stage.

## Visual Inspection

A close visual check of a metal surface can reveal tell-tale signs of corrosive attack, the most obvious of which is a corrosion deposit. These deposits are small localized discolorations on the metal surface or surface roughness.

Surfaces protected by paint or plating may only exhibit indications of more advanced corrosive attack. Such indications are the presence of blisters in the protective film, indicating that the corrosion product has a greater volume than that of the consumed metal.

In the same vein, bulges in lap joints may show a build-up of corrosion products. But, at this point, the corrosive attack is well advanced.

Sometimes the areas we wish to inspect are hidden by structural members or equipment installations, thus they are awkward to check visually. Tools such as dentists' mirrors and borescopes, can often provide the means to check an obscured area. These may avoid a time-consuming and often expensive disassembly operation.

Individual ingenuity should be encouraged if the improvised inspection methods are thorough. Magnifying glasses are valuable aids in determining whether all corrosion products have been removed during clean-up operations.

## NON-DESTRUCTIVE INSPECTION METHODS (see Chapter Fifteen)

## CORROSION APPRAISAL

When assessing the damage, it is well to consider what form of corrosive attack the metal has suffered. If the attack is uniform, often the actual damage is somewhat minor compared to the apparent damage. Conversely, localized attack, particularly pitting (which can penetrate deeply), may seem negligible at first glance, but further appraisal may show that severe damage has been sustained. It is necessary to examine each instance of corrosive attack carefully to decide whether the damage to the part is merely superficial, or if the attack has progressed further.

If corrosion has progressed beyond the superficial stage the part should be examined carefully for structural integrity. At this point corrosion damage may be considered as just another form of structural damage. The damage, if within the negligible damage limits, may be repaired by applying the instructions given in the appropriate structure repair manual.

Repairs or reinforcements not covered by the repair manual should be made only under the supervision of an experienced structures engineer.

# HYDROGEN EMBRITTLEMENT

*Hydrogen embrittlement* can either be discussed under the categories of Fatigue, Corrosion, or in a category by itself. It does include such factors as stress concentrations, foreign substances, and progressive failure. Although cyclic stress speeds up the process, the cycles are not necessary for failure.

Hydrogen embrittlement is a very complex phenomenon. We will simplify the discussion by limiting it to electroplated ferrous materials. Steel aircraft parts such as bolts, landing gear parts, and drive shafts are often electroplated with cadmium to improve the surface wear properties. We will discuss the three steps of the hydrogen embrittlement process.

## 1. Origin of Hydrogen

In the electroplating process, hydrogen is released from the electrolyte and some hydrogen atoms enter the steel material being plated. Once the plating process has been completed, the hydrogen is trapped in the steel. The atomic spacing of the plating cadmium is much smaller than that of the steel. So, the hydrogen atoms can move through the steel but cannot escape to the atmosphere through the plating.

## 2. Transport of Hydrogen

If the hydrogen influences the fracture behavior of a structure, it must be at some critical location. It can either be at this location before the metal was stressed or it must be transported to this region during the deformation process. In the plating process the hydrogen is disbursed throughout the metal, thus the method of transportation is of interest. When hydrogen originates in the metal lattice the primary transport reaction is called *lattice diffusion*. The hydrogen atoms are so small that they can move (diffuse) through the steel atoms though the metal is in the solid state. The transport of hydrogen from its equilibrium position to some critical stress location is caused by an general kinetic reaction. The hydrogen thus is concentrated at stress raisers such as crack ends. This process is shown schematically in Figure 14.9.

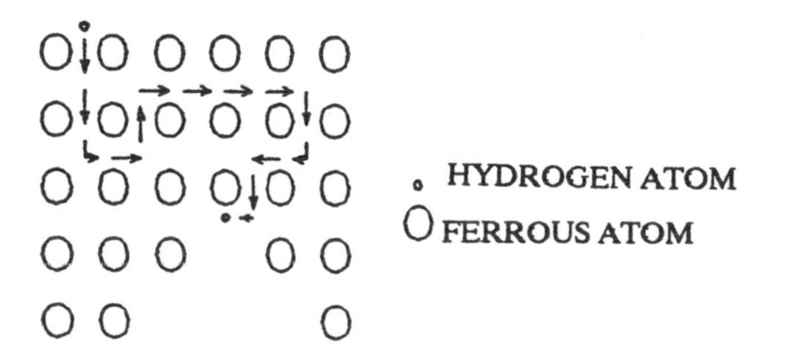


*Figure 14.9 Hydrogen atoms move by lattice diffusion.*

## 3. Hydrogen Embrittlement Interaction

Hydrogen has been observed to influence the fracture of all metals investigated to date. In steel, the hydrogen concentration at a region of high stress concentration results in precipitation of molecular (as contrasted to atomic) hydrogen.

Molecular hydrogen, ($H_2$), causes a pressure build-up at internal defects, thus it can cause failure even without applied loads. With externally applied loads the pressure build-up is much higher and failure is accelerated.

Standard procedure to protect plated aircraft parts is to bake them for about 28 hours. The higher temperature allows the hydrogen to work its way out through the plating material and escape to the atmosphere.

## REVIEW PROBLEMS

1. Name the two principal types of corrosion. Chemical + Electro chemical

2. Name two specific types of chemical corrosion. Oxidation, Acid Attack

3. Name two specific types of electrochemical corrosion.
Intergranular, Exfoliation, Stress Corrosion Cracking, Pitting, Corrosion Fatigue.

4. Corrosion has little effect on fatigue failure.

   (a) True
   (b) False

5. One good way to prevent corrosion problems on the unpainted clad aluminum alloy skin of an airplane is to keep it clean and highly polished at all times.

   (a) True
   (b) False

# CHAPTER FIFTEEN

# NONDESTRUCTIVE INSPECTION

Nondestructive testing came into use in Europe in the mid 1930's. It was used for quality control of manufactured materials and finished products. The demands of World War II and the subsequent rapid increase in technology established the need for nondestructive testing. The requirements were for greater reliability and accuracy with minimum time and cost. The aviation branches of the U.S. armed services saw that these testing methods could easily be adapted to quality control of maintenance procedures. They incorporated them into the inspection and overhaul programs and renamed them nondestructive inspection, NDI.

Nondestructive inspection is the analysis of materials without damaging or structurally altering the material. Better planning and scheduling of maintenance following systematic nondestructive inspections can reduce costly unexpected repairs. NDI can reduce the "down" time for repairs, remove hazards that could injure personnel and severely damage equipment. It and can reveal defects that would otherwise be undetected.

## THE PRINCIPAL INSPECTION METHODS

### 1. Dye Penetrants

**What can they find?**
Surface cracks and discontinuities.

**What materials can be inspected?**
Non-porous metallic and nonmetallic materials.

**How do they work?**
a. The part to be inspected must be cleaned. Paint must be removed and a degreaser applied to remove foreign materials such as grease, oil, or water. These impurities would prevent the penetrant from entering the defects.

b. The penetrant is applied to the entire surface. The penetrant is a water soluble, fluorescent liquid that will glow (usually a yellow-green color) under a black light. Small parts may be dipped into a tank of penetrant. On larger parts the penetrant can be applied by a hose or brush. It is important that all surfaces are wetted by the penetrant. After the penetrant is applied, a period of time must elapse to allow the penetrant to enter any surface defects that may be present.

Aircraft maintenance inspections are primarily concentrated on detecting fatigue cracks. Fatigue crack detection drying time for aluminum or magnesium is 30 minutes.
Penetration is shown in Figure 15.1(a).

c. The water rinse procedure removes the penetrant from the surface of the part. This rinsing must be complete and thorough so that the penetrant remaining will be only in the defects of the part. The rinsing should be done under black light so that all surface penetrant shows and can be removed.

Rinsing is shown in Figure 15.1(b).

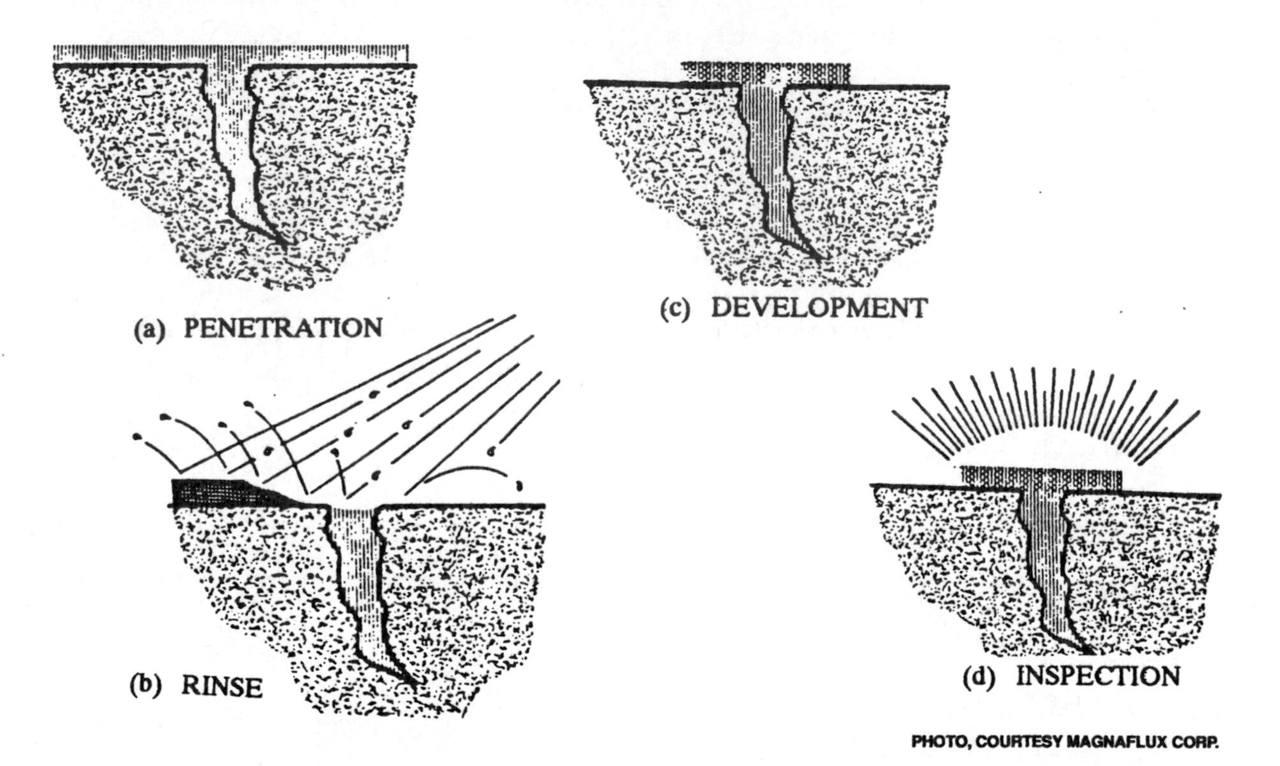


*Figure 15.1 Steps in dye penetrant inspection: (a) Penetrant, (b) Rinse, (c) Develop, (d) Inspect.*

d. After removing the surface penetrant in the rinse operation, a developer is applied to bring back to the surface any penetrant that may have found defects. This is a blotting action, and either a wet or a dry developer can be used. A drying process, at 225 degrees F, is next. In the wet developer the drying takes place after the wet developer is applied.

For the dry developer, the drying takes place before the developer is applied.
Figure 15.1(c) shows the penetrant has been drawn out of the crack by the developer.

e. The parts are inspected with a high intensity "black light" in a darkened area or booth. Cracks will show as fluorescent lines.
Figure 15.1(d) shows the inspection step of dye penetrant inspection.

**Advantages:**

1.Cheapness.
2.Portable.
3.High Sensitivity.
4.Immediate results
5.Minimum inspector skill required.

**Disadvantages:**

1.Must remove paint.
2.Can only inspect surface flaws.
3.No permanent inspection results.
4.Direct visual detection required.
5.Requires high degree of cleanliness.

## 2. Magnetic Particle Inspection

**What can it find?**
Surface cracks and sub-surface defects close to the surface.

**What materials can be inspected?**
Magnetic materials only.

**How does it work?**
a. The parts to be inspected should be cleaned prior to inspection. It is desirable, although not practical, to chemically clean the part. But removal of gross contamination is required to provide a surface that can be inspected. Paint has the effect of converting surface flaws into sub-surface flaws. In deciding whether paint should be removed, the relative thickness of the paint and the size of the smallest flaw being sought must be considered. Plating is usually thinner than paint and up to .004 inches thickness will not interfere with inspection.

b. A magnetic field is induced into any ferromagnetic part by either longitudinal magnetization or circular magnetization.

Longitudinal magnetization is done by placing the part in a coil carrying electrical current. Cracks in the lateral direction are found using this method. This is shown in Figure 15.2(a).

Circular magnetization is done by passing a current lengthwise through the part to be inspected. This will reveal cracks in the longitudinal direction. This is shown in Figure 15.2(b).

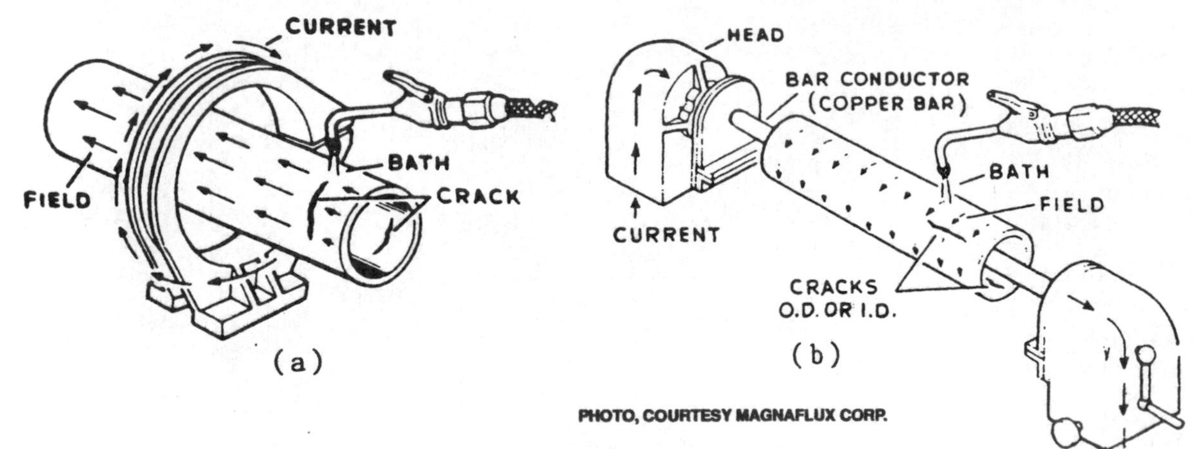


PHOTO, COURTESY MAGNAFLUX CORP.

*Figure 15.2 (a) Longitudinal, (b) Circular magnetization.*

c. There are two principal ways of applying magnetic particles to the surface. The wet method, most often used in the NDI laboratory, has fluorescent magnetic particles suspended in an oil or water bath. The wet method particles are the most desirable for the detection of fatigue cracks. The wet inspection bath is applied to parts from a hose and nozzle (see Figure 15.2) or by dipping parts into it. This method is most effective with stationary equipment.

Wet magnetic particle inspection can be conducted in the field using portable equipment. This equipment is shown in Figure 15.3. The "bath" comes in push-button spray cans containing red colored magnetic particles suspended in oil.

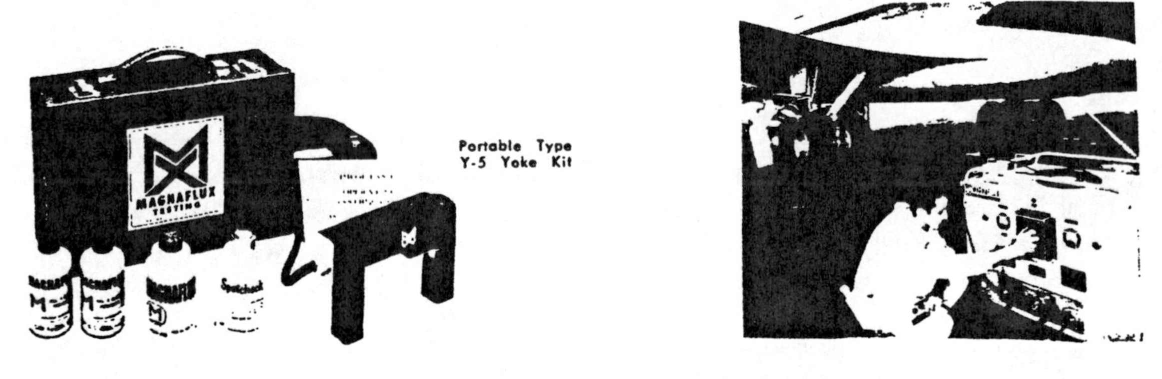


PHOTO, COURTESY MAGNAFLUX CORP.

*Figure 15.3 Portable magnetic particle inspection equipment.*

The dry method is not as accurate as the wet method but may be preferred for field use. In this method the magnetic particles are in a finely dispersed powder form and obtain their mobility by being suspended in air.

The field methods are used on large parts, such as landing gears, which are not easily removed for inspection in the shop.

d. Examination of parts coated with fluorescent particles used in the wet bath procedure is done after the part is allowed to drain for a few seconds.

The fluorescent material will show up when viewed under an ultraviolet "black light" in a darkened area. Cracks will show up as seen in Figure 15.4.

Inspection of non-fluorescent colored particles used in field methods can be made under normal white light.

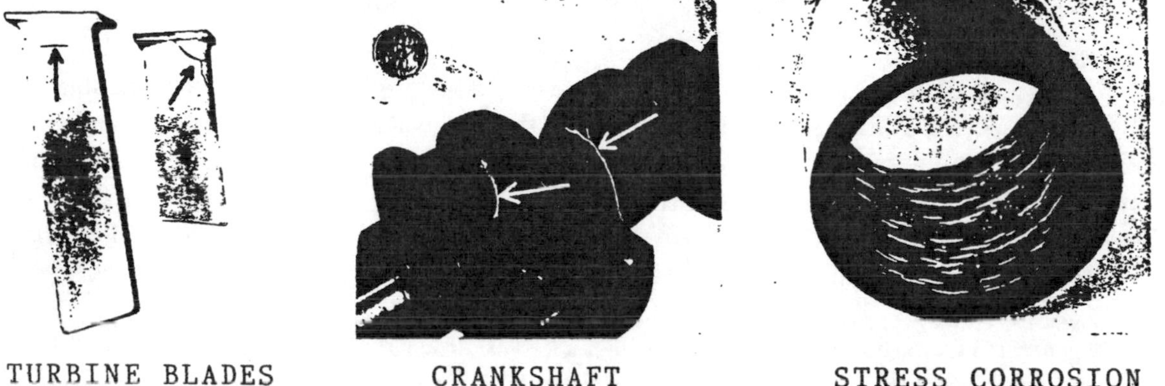


PHOTO, COURTESY MAGNAFLUX CORP.

*Figure 15.4 Cracks shown by magnetic particle inspection.*

e. Demagnetization of aircraft parts is required to prevent the part from affecting magnetic compasses and other equipment.

The part can also attract loose steel scrap material if it is not demagnetized.

**Advantages:**

1.Semi-portable.
2.Sensitive to small flaws.
3.Finds sub surface defects (if close to surface)
4.Moderate inspector skill required

**Disadvantages:**

1.Only works on magnetic materials
2.Removal of most surface coatings required
3.No permanent test results
4.Must demagnetize the part after inspection

## 3.Radiography (X-Rays and Gamma Rays)

**What can they find**

Internal flaws (with some depth limitations).

**What materials can be inspected**

All metallic and non-metallic materials.

**How do they work**

A *radiograph* is a "shadow" picture produced by the passage of x-ray or gamma rays through an object onto a film. Radiation proceeds from its source in straight lines to the specimen under examination. Some rays pass through while others are absorbed. The amount of rays passing through depends upon the energy of the radiation and the nature and thickness of the material. If the specimen contains a void, for instance, more radiation will pass through the void than through the surrounding material. This will produce a dark spot on the developed film, corresponding to the shape and position of the void.

*X-rays* are generated from the sudden stopping of high-speed electrons in an evacuated x-ray tube. When the high-speed electrons are stopped their kinetic energy transforms into both x-rays and heat. Increasing the voltage will increase the number of electrons emitted per unit of time resulting in a more intense beam.

*Gamma rays* are emitted from the disintegrating nuclei of natural elements or artificial radioactive isotopes. Although very similar to x-rays, gamma rays frequently have a much shorter wave length and a greater penetrating power.

The steps in making a radiograph are few and appear somewhat simple, but certain decisions must be made at each step in the process that are of extreme importance in deciding the validity of the test. They are:

a. The specimen is placed at the desired angle.
b. The films, screens, and a radiation source are selected.
c. The proper source-to-film distance is selected.
d. Scattered radiation is controlled.
e. The exposure time is calculated.
f. The film is exposed, developed, and inspected.

A great amount of skill is required for successful radiography, not only in the actual exposing of the film but also in film interpretation. The interpreter must be able to differentiate slight changes of gray color.

He must be thoroughly familiar with the shape of the specimen, the portion shown on the film, and the material of the specimen. If the interpreter is not the same individual who exposed the film, he must have exact knowledge of the techniques that were used. There should be adequate records accompanying the film that give the radiation source used, the direction of the beam, the section radiographed, and other pertinent information. The interpreter must then make judgments and comparisons of what he sees to what the standards and specifications are.

*Fluoroscopy* is similar to x-ray except that no film is used. The x-rays pass through the object being inspected and illuminate a fluorescent screen. Since no film is used this process is a low-cost, quick method of internal inspection. The object is viewed by the operator and immediately accepted or rejected. If a part is found to contain a flaw and is rejected, a photograph can be taken as proof.

**Advantages:**

1.Can detect surface and sub-surface defects
2.A permanent record is retained
3.Minimum part preparation required

**Disadvantages:**

1.Most expensive.
2.Radiation dangers to personnel are high
3.Method is highly directional, may need exposures from several different angles.
4.High degree of interpretive skill required.

## 4. Ultrasonic Inspection

**What can it find?**
External and internal flaws(no depth limitations)
Thickness dimensions of materials
Leaks in pneumatic systems

**What materials can be inspected?**
All metallic and non-metallic materials

**How does it work?**
*Ultrasonic inspection* makes use of high-frequency mechanical vibrations. These vibrations are similar to sound waves but far beyond the audible range. They are waves created by particle vibration. They do not travel well through air but travel easily through solids or liquids. Among the characteristics of these ultrasonic vibrations is that they will be reflected at discontinuities or boundaries of different elastic or physical properties.
This "echoing" characteristic is the basis for the most commonly used ultrasonic detection method, the *pulse-echo technique.*

The heart of the pulse-echo technique is the method of transforming electrical pulses into mechanical vibrations, and transforming the mechanical vibrations back into electrical pulses. This transformation is made by a device known as a transducer. It works on the principle that if an electrical pressure (voltage) is applied across a crystal, the crystal thickness will vary as the frequency of the applied voltage.

Conversely, if mechanical pressure is exerted on the face of the crystal, it will generate a small voltage of the same frequency as the applied mechanical vibration.

The ultrasonic pulse is generated by producing a radio frequency pulse of the desired frequency at a precise time, then converting the pulse into ultrasonic vibrations by the piezoelectric crystal transducer. The crystal is actuated for a controllable period (about 2 millionths of a second) resulting in a short pulse of sound waves.

The pulse travels through the material to the opposite boundary where it is reflected back to the source as an echo. After the crystal has given off this short burst of vibrations, it stops vibrating for a period long enough to receive the returning echoes. This cycle of transmitting and receiving is repeated at a rate of 60 or more times per second, according to the type of ultrasonic equipment used. The ultrasonic waves can be directed through the material to be inspected in several different ways, some of which are shown in Figure 15.5.

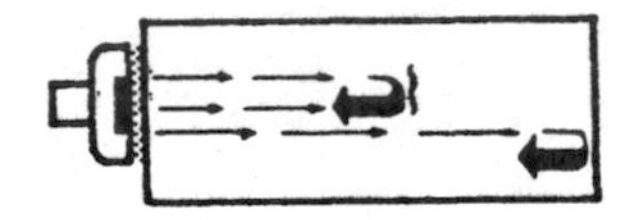

**Longitudinal Wave** — transmitted perpendicularly through the test surface. The beam reflects from flaws, and the opposite side.

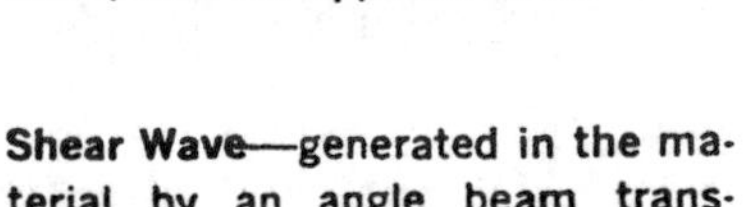

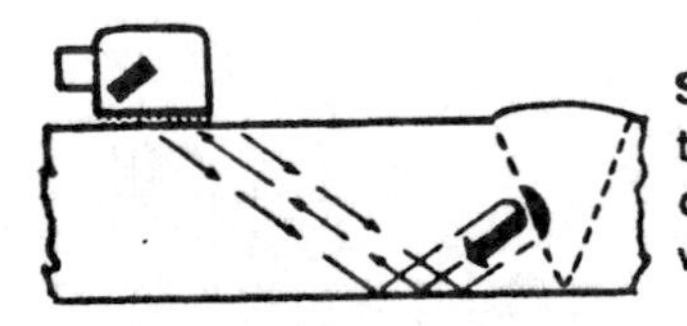

**Shear Wave**—generated in the material by an angle beam transducer: used to test materials and welds for sub-surface defects.

**Surface Wave**—travels along, and slightly below the surface of the material. Pulses will travel over irregularly shaped surfaces.

**Lamb Wave** — generated by angle beam transducer: travels thru thin, parallel sided material, reflecting from defects.

PHOTO, COURTESY MAGNAFLUX CORP.

*Figure 15.5 Types of pulse echo ultrasonic waves.*

Reflection of the ultrasonic vibrations will occur at the boundary between two different materials. The ultrasonic beam can be compared to a beam of light traveling through space and being reflected by many mirrors. The path traveled on the return of the beam to its source, depends upon the angles at which it impinges upon the reflecting surfaces, and the number and locations of these surfaces.

The detection of an internal crack can be described by considering the case where longitudinal waves are transmitted through the inspected part as shown in Figure 15.6. The operator observes a cathode ray tube CRT and sees three pips. The left pip is the echo from the face of the part, the second pip shows the crack and the third pip shows the rear face of the part. Distances between pips are proportional to the thickness of the test part.

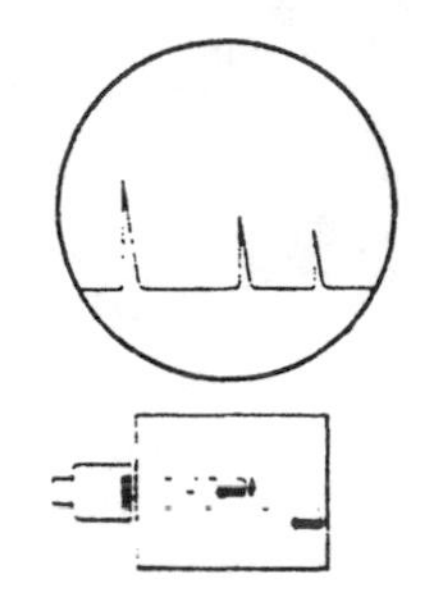

***Figure 15.6 Indication of a crack by ultrasonic inspection.***

Thickness of pipe, tubing, sheet, strip, plate or other test items can be measured by ultrasonic instruments. The instrument is calibrated using a standard depth test block and then shows thickness to .001 inch.

Thickness measurement is shown in Figure 15.7.

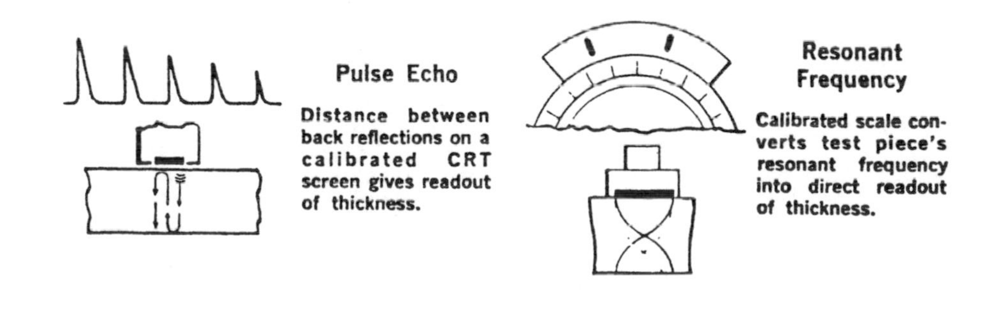

***Figure 15.7 Thickness measurement by ultrasonics.***

Ultrasonic equipment can find delaminations, corrosion, and fire damage, all of which change the thickness of the material.

Figure 15.8 shows inspection of a pipe for internal corrosion.

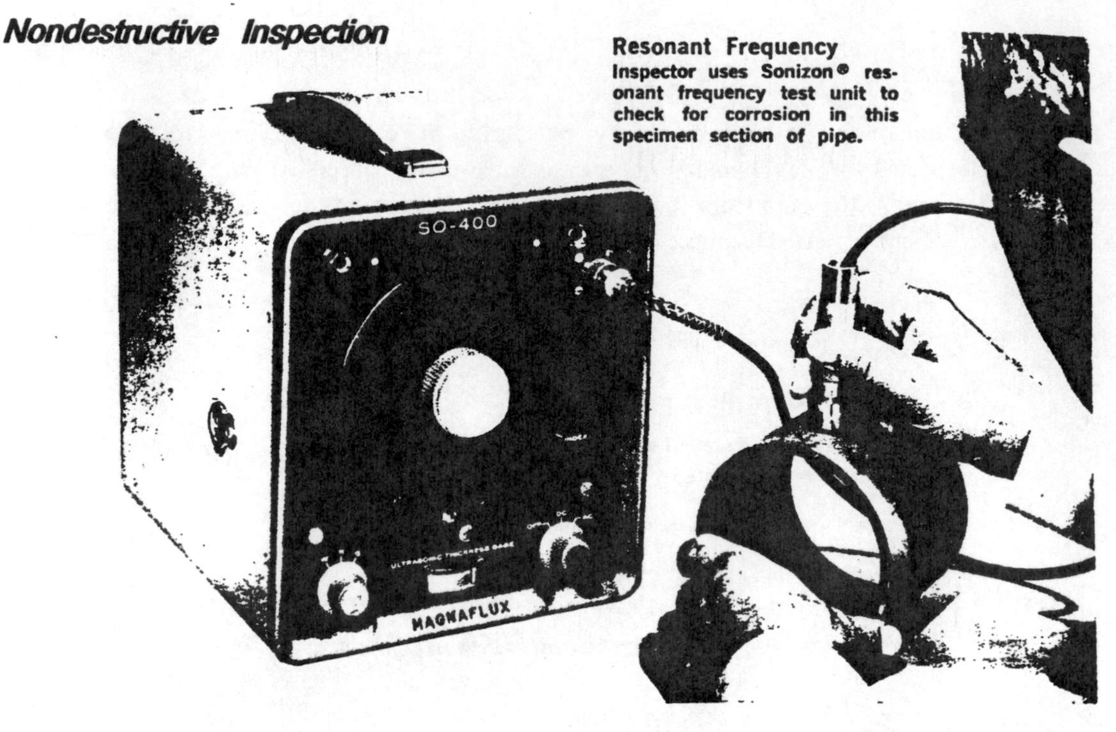


PHOTO, COURTESY MAGNAFLUX CORP.

*Figure 15.8 Corrosion detection by ultrasonics.*

A special microphone type probe can be used to amplify sounds such as escaping air from pressurized pneumatic systems. This is useful in finding small leaks in the system.

**Advantages:**
1.Good for surface and sub-surface defects.
2.Sensitive to small defects.
3.Immediate inspection results.
4.Little part preparation.
5.Wide range of material thickness can be inspected.

**Disadvantages:**
1.No permanent record of flaw.
2.Inspection is directional depends on orientation of flaw.
3.Highly skilled inspector required.

## 5. Eddy Current Inspection

**What can it find (or do)?**

Find flaws (cracks, voids, inclusions, seams, laps)
Sort parts according to alloy, temper, conductivity, and other metallurgical factors
Gauge metals according to size, shape, and thickness

**What materials can be inspected?**
All structural metal parts

**How does it work?**
An *eddy current* is an induced electrical current circulating within a mass of metal. In this technique a coil carrying alternating current induces an alternating current (eddy current) in the part being inspected. If a flaw is present in the part the eddy current will be compressed and distorted. This will result in a deflection of the needle of the test meter. The eddy current method does not reveal the nature of the flaw but merely locates it. The probe type of coil is used in inspection procedures. This is a small coil that can be placed on the surface of the inspected part so that the axis of the coil is perpendicular to the surface. This is shown in Figure 15.9(a). The use of a probe coil in the inspection of rivet holes for possible fatigue cracks is shown in Figure 15.9(b).

PHOTO, COURTESY MAGNAFLUX CORP.

*Figure 15.9 (a) Probe coil; (b) Inspecting rivet holes.*

The "through coil" is used in production line quality control. The major advantage of this type is that mechanization is easy and the speed of inspection can be high. It is particularly suited to the inspection of cylindrical objects such as bar stock and tubing. Figure 15.10(a) shows the principle of the through coil and 15.10(b) shows a typical production line configuration. The test part is moved through the coil and any flaw in the test part will cause an unbalanced eddy current between the two secondary coils. This will trigger a read-out device such as a spray paint gun that then marks the defect. Other read-out devices are meters, scopes, pen recorders, flashing lights, audible alarms, counters etc.

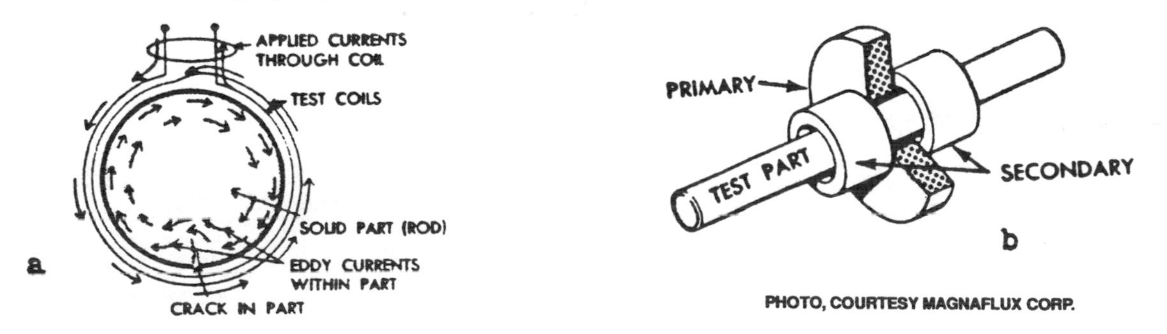


PHOTO, COURTESY MAGNAFLUX CORP.

*Figure 15.10 Through coils.*

*Nondestructive Inspection*

**Advantages:**

1.Portable.
2.Moderate cost.
3.Immediate results.
4.Sensitive to small flaws.
5.Little part preparation.

**Disadvantages:**

1.Surface must be accessible to probe.
2.Rough surfaces interfere with sensitivity.
3.Suitable for metals only.
4.No permanent record.
5.Highly skilled inspector required.
6.Time consuming to scan large areas.
7.Difficult to use on parts that have varying cross sections

## REVIEW PROBLEMS

Which non-destructive inspection method would be most desired for each of the following?

1. Fatigue crack in the unpainted skin of an aircraft:
   (a) ultrasonic; (b) magnetic particle; (c) dye penetrant; (d) radiographic

2. Crack in an automobile crank shaft:
   (a) ultrasonic; (b) magnetic particle; (c) dye penetrant; (d) radiographic

3. Water in a honeycomb tail surface:
   (a) ultrasonic; (b) magnetic particle; (c) dye penetrant; (d) radiographic

4. Pneumatic system leak:
   (a) ultrasonic; (b) magnetic particle; (c) dye penetrant; (d) radiographic

5. Sub-surface crack in magnesium wheel casting:
   (a) eddy current; (b) magnetic particle; (c) dye penetrant; (d) radiographic

# ANSWERS TO PROBLEMS

## CHAPTER ONE

1. (a); 2. (d); 3. (c); 4. (c); 5. (b);
6. (c); 7. (c); 8. (c); 9. (c); 10. (d).

## CHAPTER TWO

1. 1.56 G, 11,913 pounds; 2. 170 knots; 3. Plane B;

4. Plane A; 5. 7 Gs:

## CHAPTER THREE

1. 7.33 Gs; 2. 1.3M; 3. No; 4. 6.2 Gs; 5. 0.55M.

## CHAPTER FOUR

1. (c); 2. (b); 3. (a); 4. (c); 5. (a).

## CHAPTER FIVE

1. (c); 2. (c); 3. (b); 4. (b); 5. (d);

6. (a); 7. Shear on 45°; 8. Tension on 90°.

## CHAPTER SIX

1. (c); 2. (b); 3. 69,444 psi; 4. 25,465 psi.

## CHAPTER SEVEN

1. (a); 2. (a); 3. (b); 4. (c); 5. (a).

## CHAPTER EIGHT

1. psi, in/in; 2. 0.1 in/in; 3. 12 in; 4. (c);

5. 30,000,000 psi, steel; 6. 8 in; 7. D; 8. L; 9. J; 10. B.

## CHAPTER NINE

1. T; 2. F; 3. F; 4. Slope of tangent to f-e diagram. To find $f_c$ for short column; 5. 1.75 times; 6. Compression;
7. (a) 0.288675 (b) 20 in; (c) 69.281465; (d) 20,562 psi.

## CHAPTER TEN

1. (c); 2. (c); 3. (b); 4. (a); 5. (d); 6. (b); 7. d.

## CHAPTER ELEVEN

1. (c); 2. Mosquito, Spruce Goose; 3. (a); 4. (c);

5. (a); 6. (b); 7. (b); 8. (b).

## CHAPTER TWELVE

1. (a); 2. (c); 3. (d); 4. (c); 5. (b); 6.(b).

## CHAPTER THIRTEEN

1. (b); 2. (b); 3. (b); 4. (b); 5. (d); 6. (a)

## CHAPTER FOURTEEN

1. Chemical and Electrochemical

2. Oxidation, Acid attack

3. Intergranular, Exfoliation, Stress Corrosion cracking, Pitting, Corrosion Fatigue

4. (b)

5. (b)

## CHAPTER FIFTEEN

1. (c); 2. (b); 3. (d); 4. (a); 5. (a).

# APPENDIX

This book was written to serve as a useful tool in the understanding and recognition of failures in aircraft parts and materials. An important part of the work of an aircraft mishap investigator is to examine these failures to determine what happened, how it happened and why it happened.

In this appendix we will summarize and show more examples of parts and materials which have failed. The failure may be a result of errors in design, manufacture, testing, maintenance, inspection or operational procedures.

Identification of the type of failure is not the final job of the investigator. The cause of the mishap must be determined. The type of failure is, however, an excellent clue to this determination.

Many failures require extensive technical examination, far beyond the capabilities of the investigator. For instance, finding that corrosion is a factor is probably within the ability of the investigator, but determining the type of corrosion may require examination by experts in this field.

## TENSION LOADING FAILURES

In Chapter Five we stated that the principal stress resulting from a tension load is tension stress and that it is a maximum on a plane that is 90° to the applied tension load. We also learned that there is also a component shear stress which is a maximum on the 45° plane. The maximum value of the shear stress is one-half that of the maximum tension stress. From this we concluded that brittle materials, which are relatively weak in tension, would fail in tension on the 90° plane. This is shown in Figure A.1.

Extremely ductile materials are relatively weak in shear and thus excessive tension loads result in shear failures on the 45° plane. Such failures are shown in Figure A.2.

Most metals used in aircraft are relatively ductile but not to the extent of those shown in Figure A.2. Upon failure under tension loads they exhibit a combination of brittle and ductile failure as shown in Figure A.3. The center portion of a cylindrical portion shows 90° brittle fracture while the sides show 45° "shear lips" failure. Considerable necking down is also evident.

# TENSION LOAD - TENSION FAILURE

## EXTREMELY BRITTLE METAL

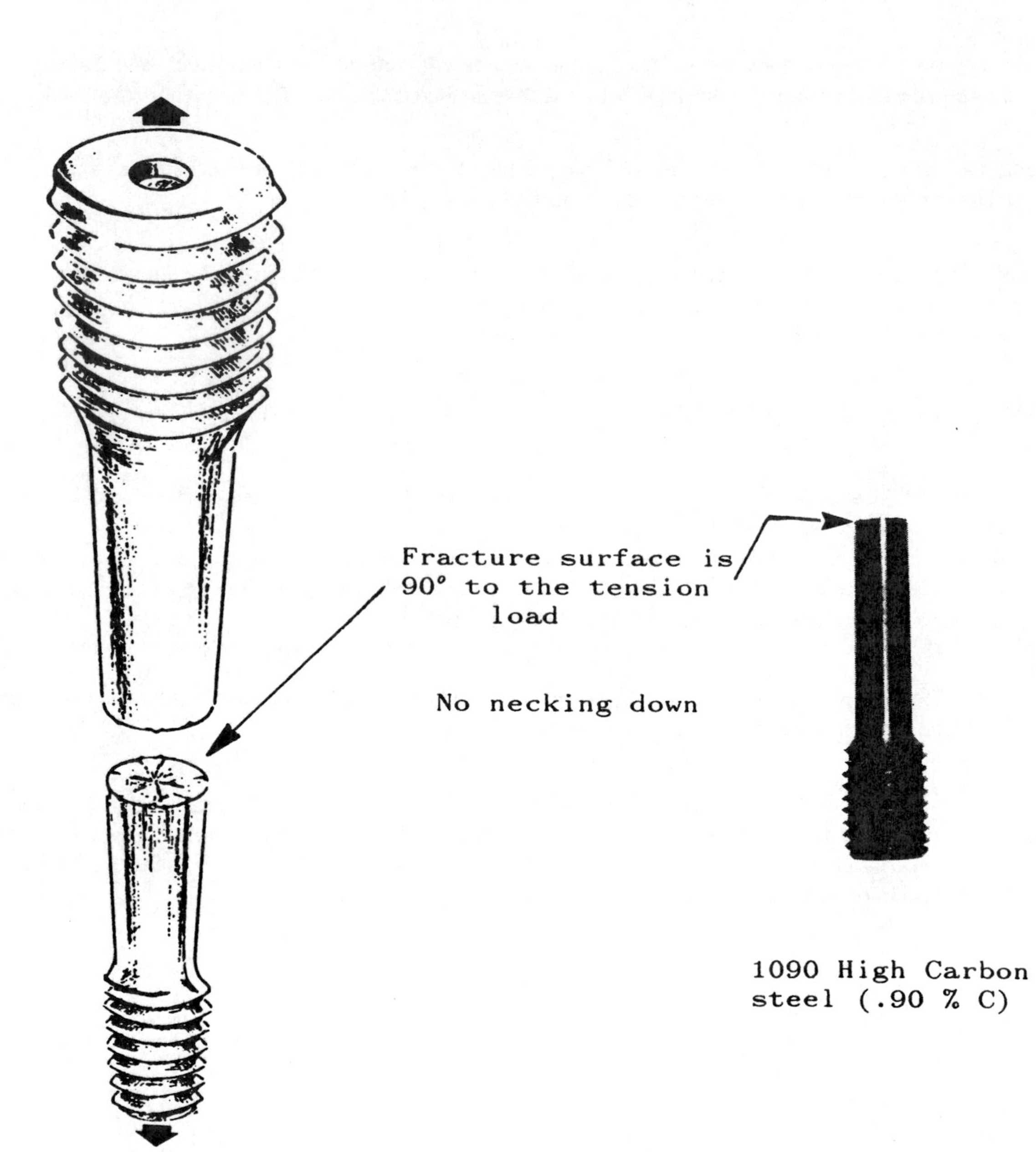


*Figure A.1*

# TENSION LOAD - SHEAR FAILURE

## EXTREMELY DUCTILE METAL

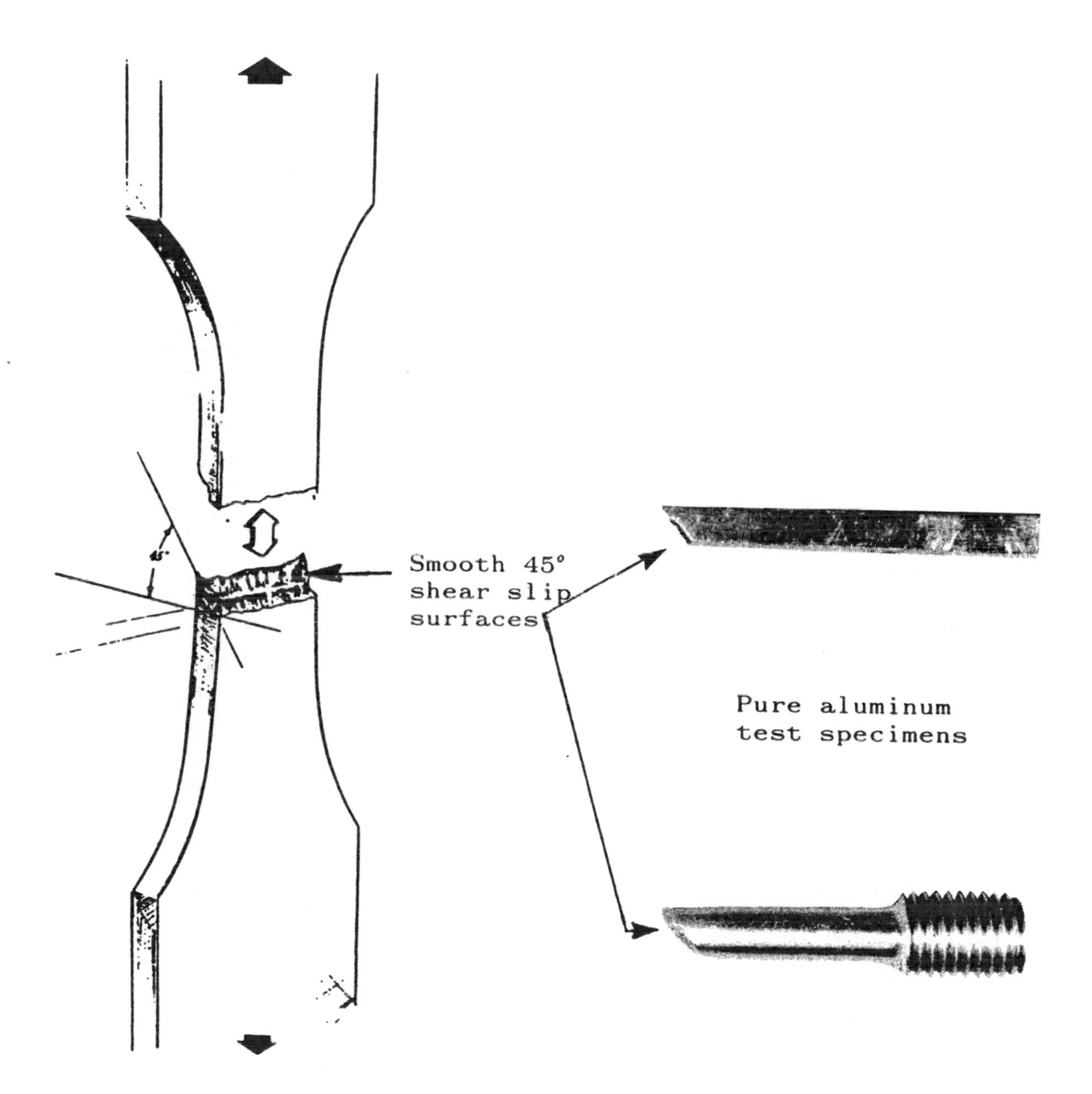

*Figure A.2*

# TENSION LOAD - COMBINATION FAILURE

## MODERATELY DUCTILE METAL

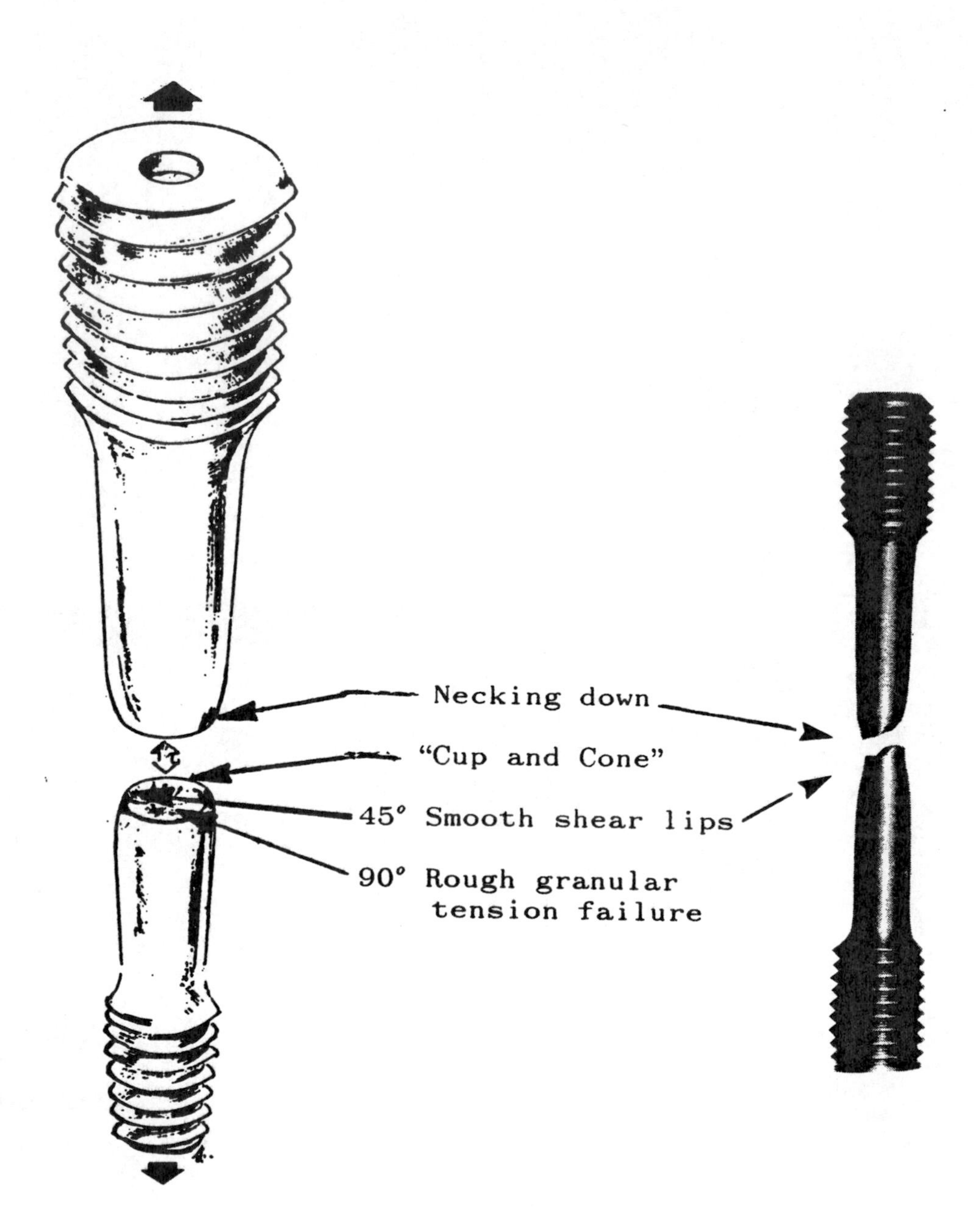

*Figure A.3*

# TENSION LOAD - COMBINATION FAILURE

## MODERATELY DUCTILE MATERIAL (con'd)

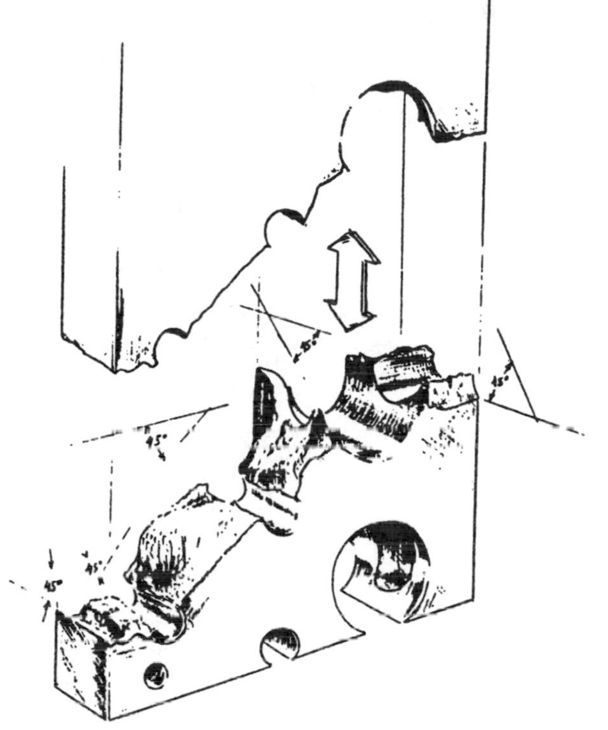


2024 Aluminum Alloy

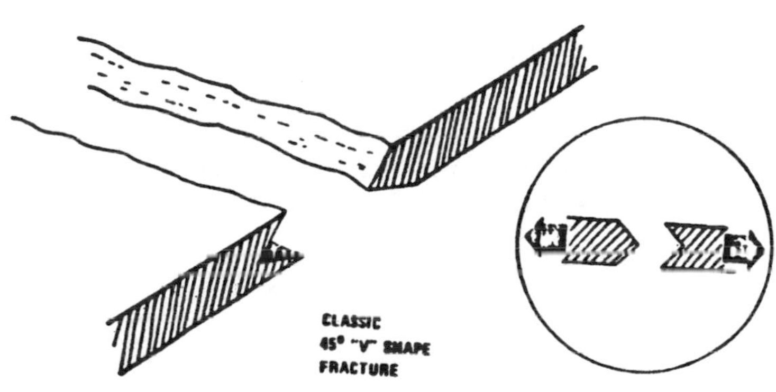


7075 Aluminum Alloy

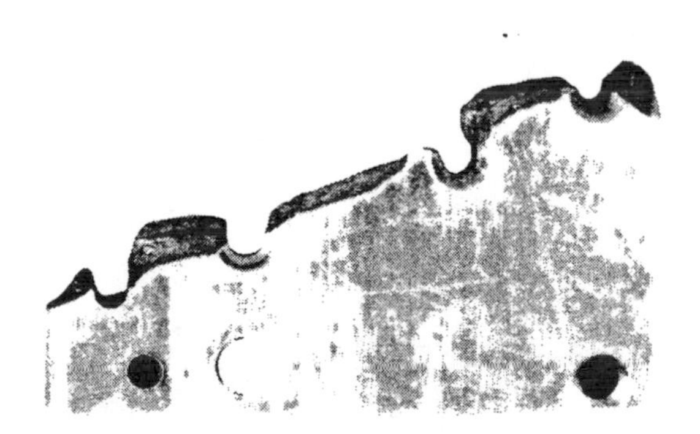

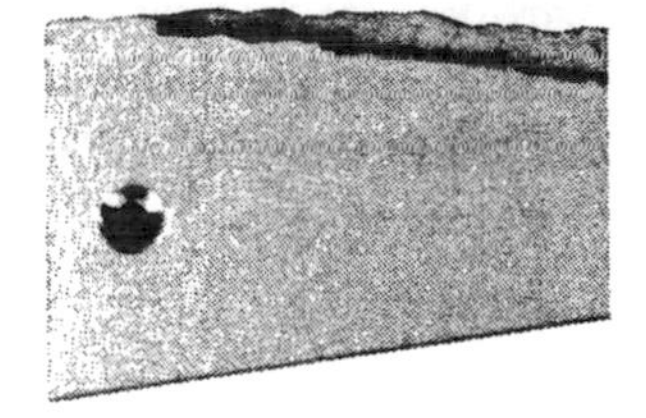

*Figure A.4*

## COMPRESSION LOADING FAILURES

In Chapter Five we discussed compression loads and resulting compression stresses. We stated that compression stress, $f_c$, was calculated by the formula $f_c = P/A$. If a short solid ductile metal part is loaded in compression, and the compression stress causes the material to deform excessively so that the part can no longer perform its mission, the part is said to have failed. This type of failure is not common in aircraft parts.

A second type of failure under compression loads is a shear failure. This type of failure was also discussed in Chapter Five under the "Stresses on Oblique Sections" section. Here it was stated that shear stresses were developed as a result of compressive loads and that the shear stresses were a maximum on the 45° plane. Shear failures resulting from compressive loads are found in brittle metals. This is in direct contrast to shear failures resulting from tension loads which occur in ductile metals. Some aircraft metals, such as high strength aluminum alloy 7075, exhibit brittle characteristics and fail in shear. An example of such failure is seen in Figure A.5. A wood specimen, although not common to modern aircraft, which failed in shear under compressive loading is also shown in this figure.

Buckling is the most common type of failure of aircraft parts subjected to compression. This was discussed in detail in Chapter Nine. Examples of buckling failures are shown in Figures A.6, A.7, and A.8.

## SHEAR LOADING FAILURES

Shear loads produce shear stresses, compression stresses and tension stresses. Each of these stresses have the same value but act upon different planes relative to the applied shear loads. This was discussed in Chapter Five under the heading of "Shear Loads."

Illustrations of shear failures resulting from shear loads are shown in Figure A.9. This type of failure is common in ductile solid bars, drive shafts, and rivets.

Illustrations showing compression (buckling) failures resulting from shear loads are shown in Figure A.10. This type of failure is common in thin sheets or plates of ductile materials.

Illustrations of tension failures resulting from shear loads are shown in Figure A.11. This type of failure is rare in aircraft as it occurs only in brittle materials, not commonly used in aircraft. The tension cracks in this figure appeared after the buckling occurred. Thus the primary failure was buckling.

# COMPRESSION LOAD - SHEAR FAILURE

## 7075 AL ALLOY and WOOD SPECIMENS

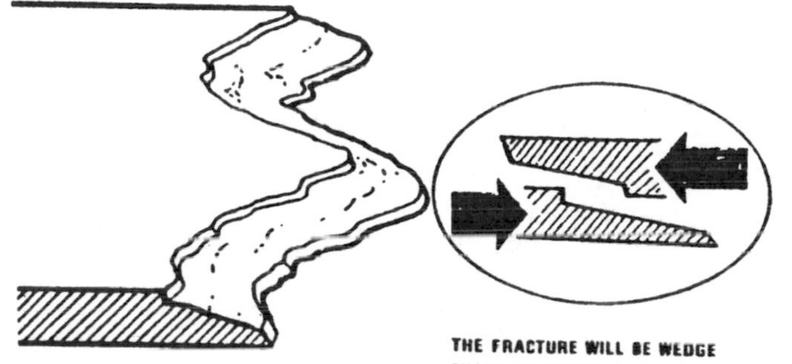


7075 Aluminum Alloy

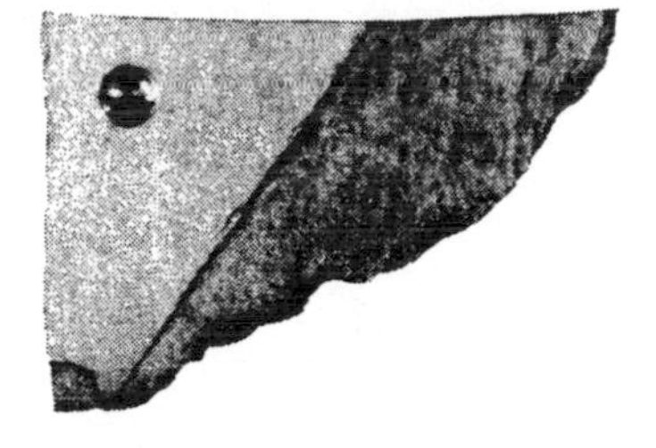

Wood

***Figure A.5***

# COMPRESSION LOAD - BUCKLING FAILURE

## DUCTILE METAL

LONG COLUMNS - Failure occurs in the elastic range.

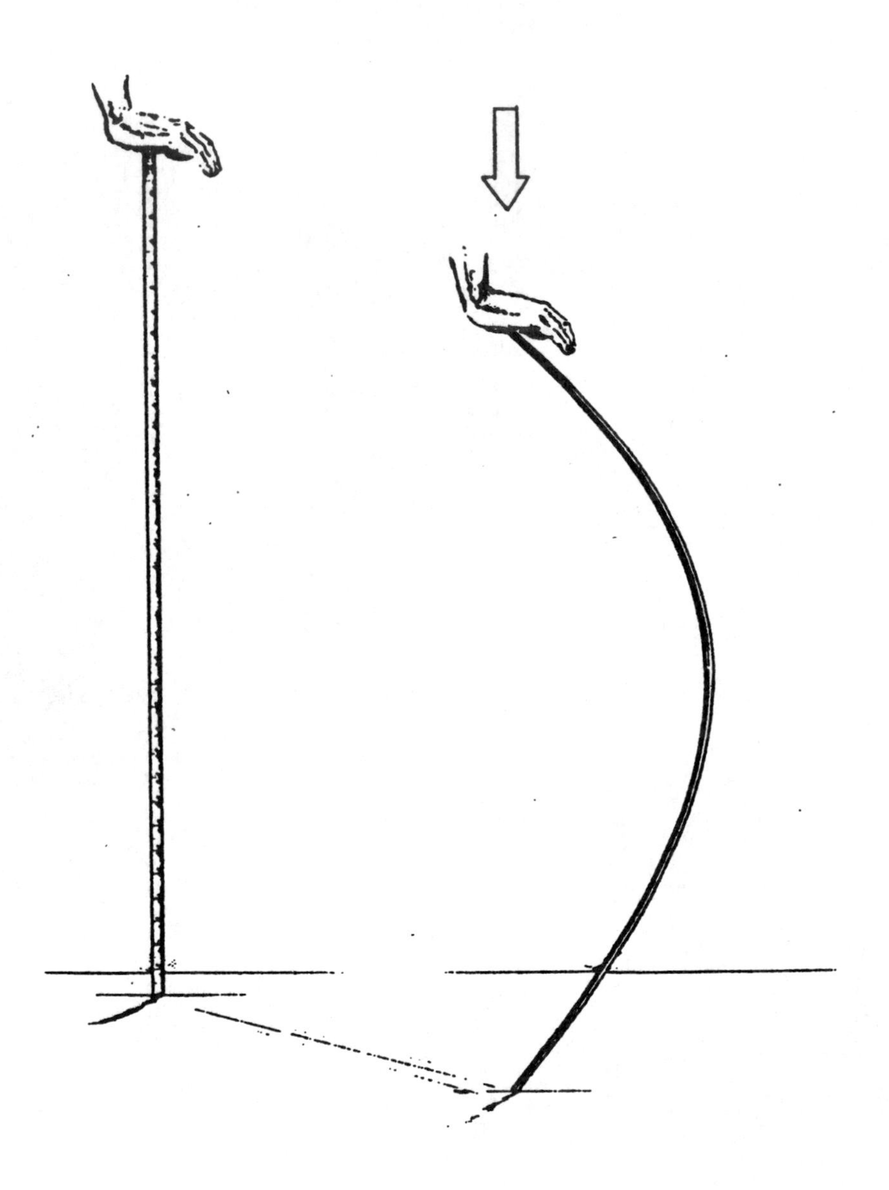

***Figure A.6***

# COMPRESSION LOAD - BUCKLING FAILURE

## DUCTILE METAL (con'd)

SHORT COLUMNS - Failure occurs in the plastic range.

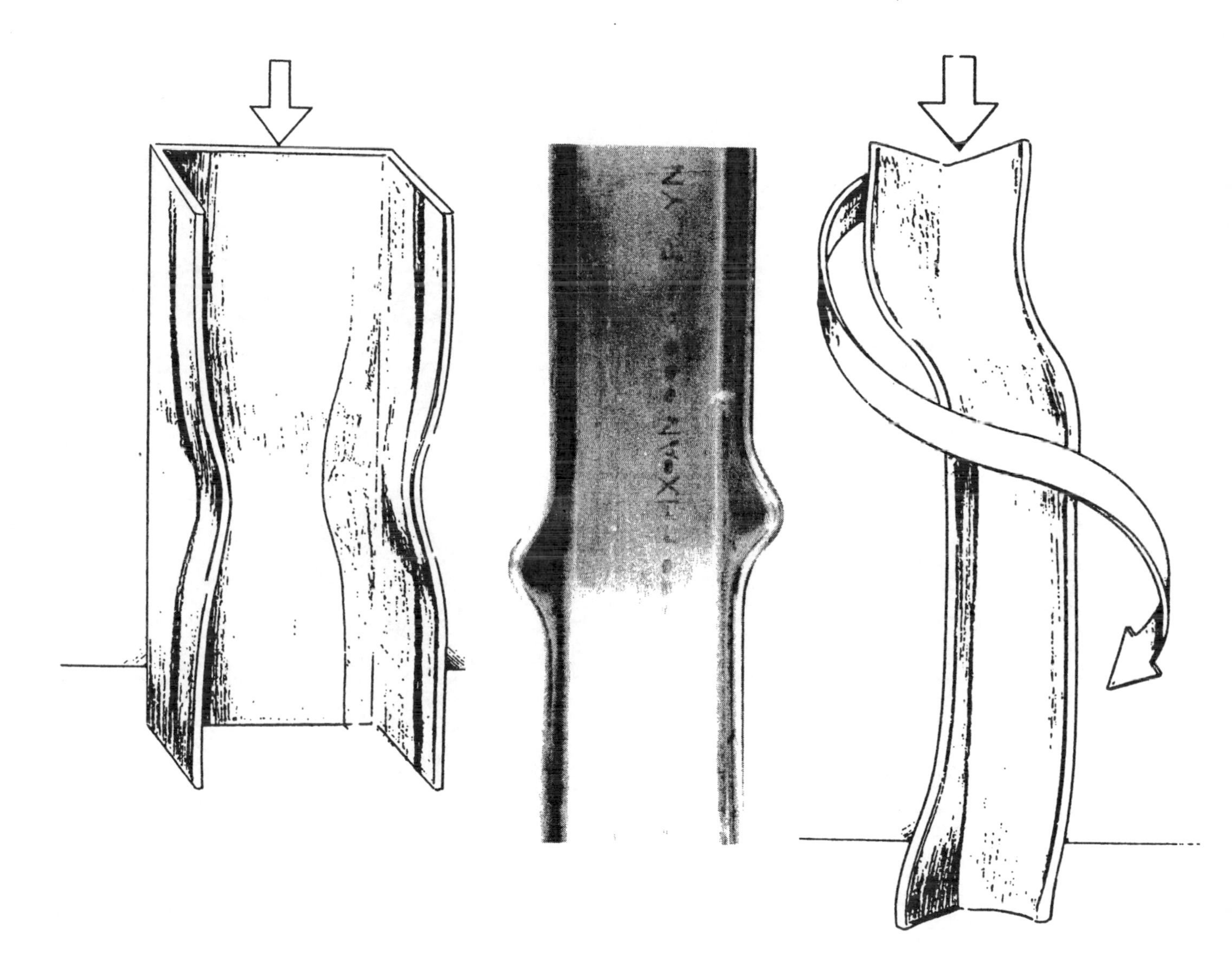

***Figure A.7***

# COMPRESSION LOAD - BUCKLING FAILURE

## DUCTILE METAL (con'd)

ROUND TUBES - Diamond shaped plastic buckles.

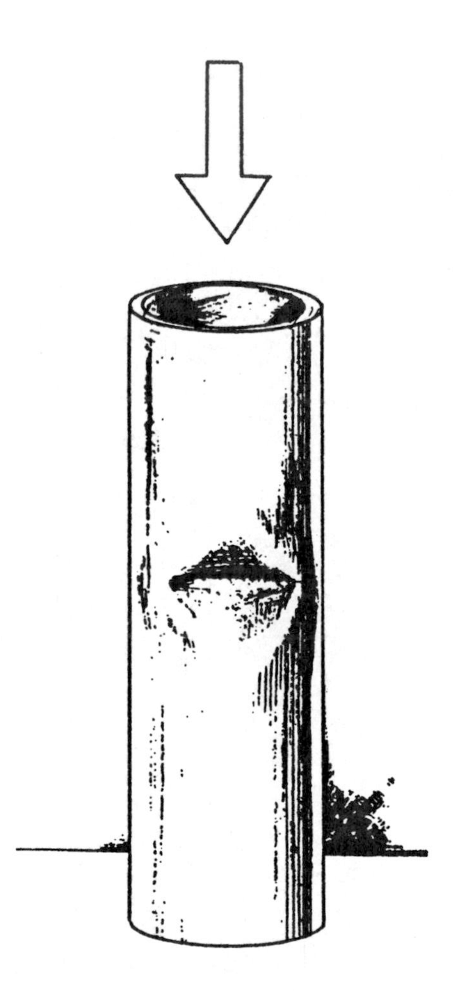

***Figure A.8***

# SHEAR LOAD - SHEAR FAILURE

## DUCTILE METAL

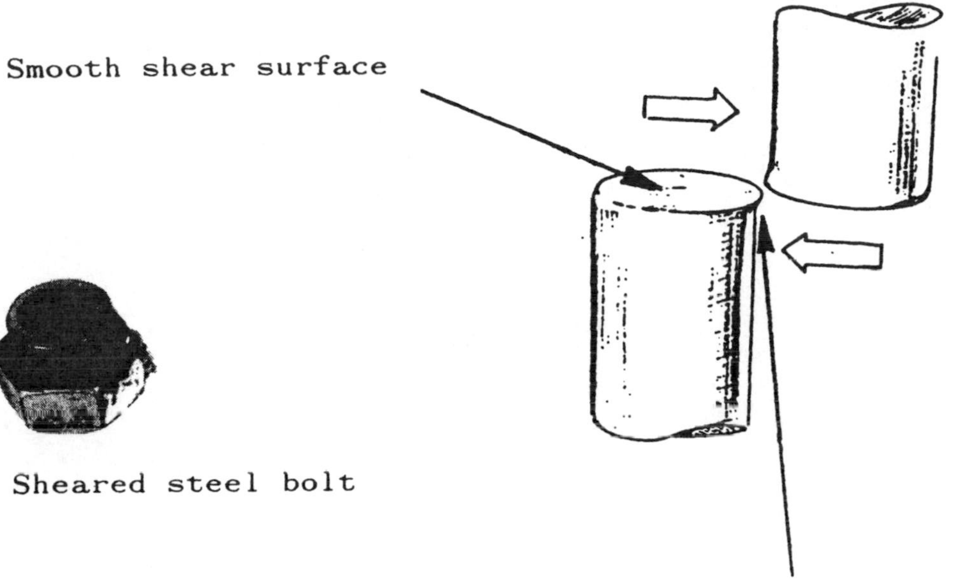

Sheared steel bolt

BRITTLE METAL

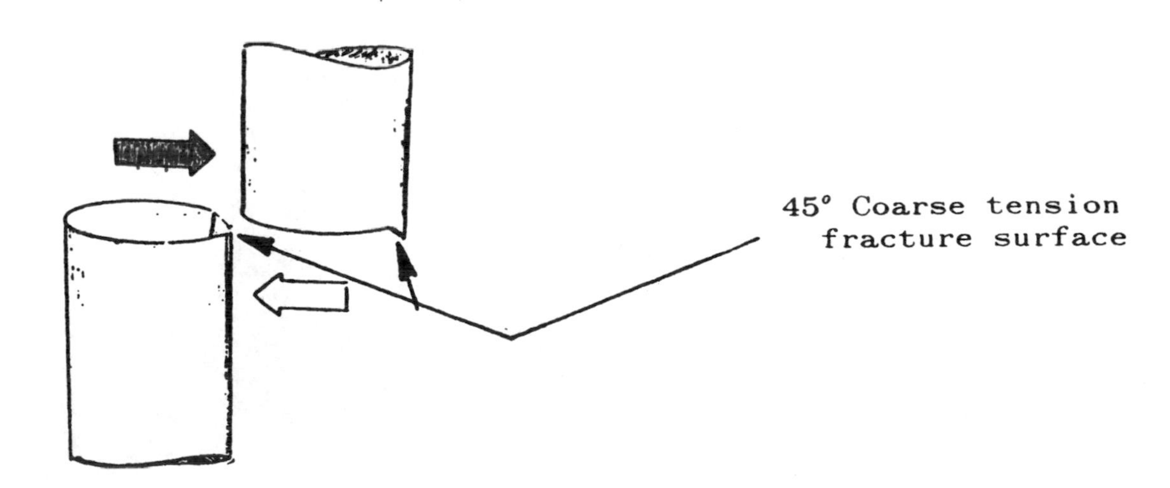

*Figure A.9*

# SHEAR LOAD - SHEAR FAILURE

## DUCTILE METAL (con'd)

Steel "Wrist Pin"

Shear has started but not completed

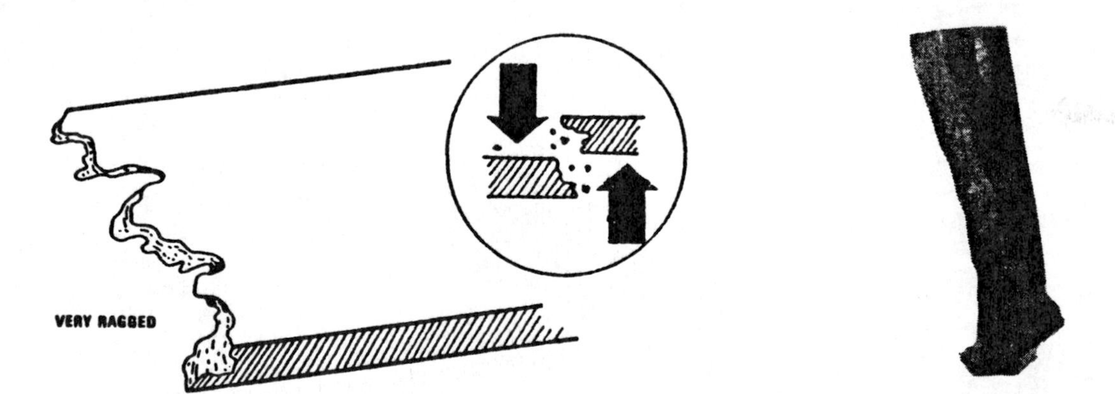


7075 Aluminum Alloy Sheet

*Figure A.10*

# SHEAR LOAD - BUCKLING FAILURE

## DUCTILE METAL PANEL

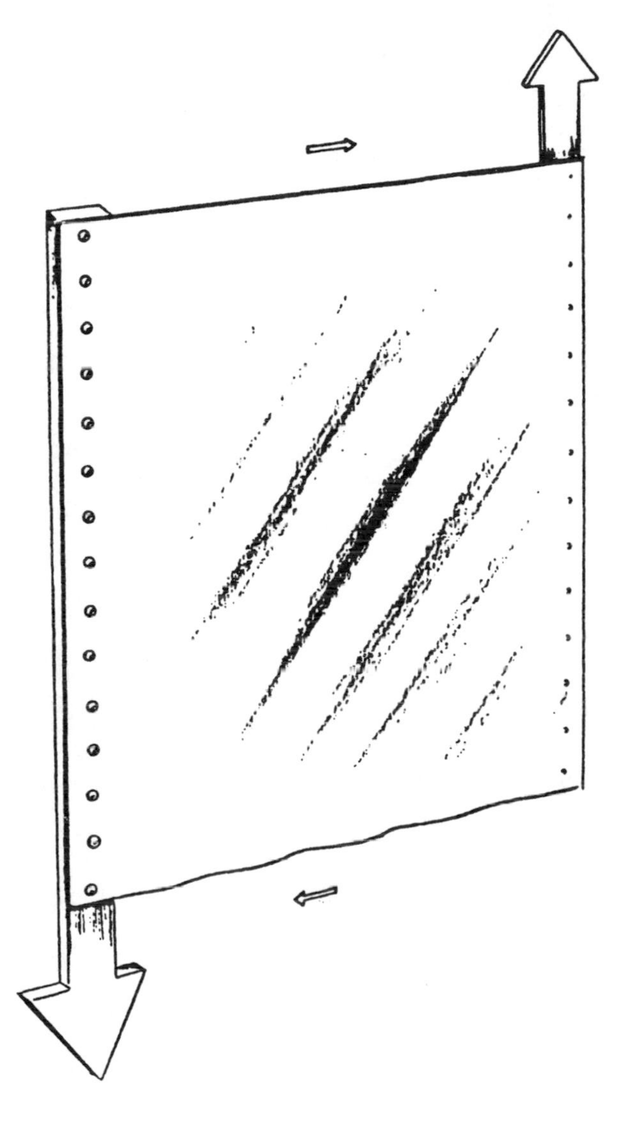

NOTE: A line drawn between the large arrow heads shows the direction of the buckles.

***Figure A.11***

# SHEAR LOAD - TENSION FAILURE

## DUCTILE PANEL - SECONDARY FAILURE

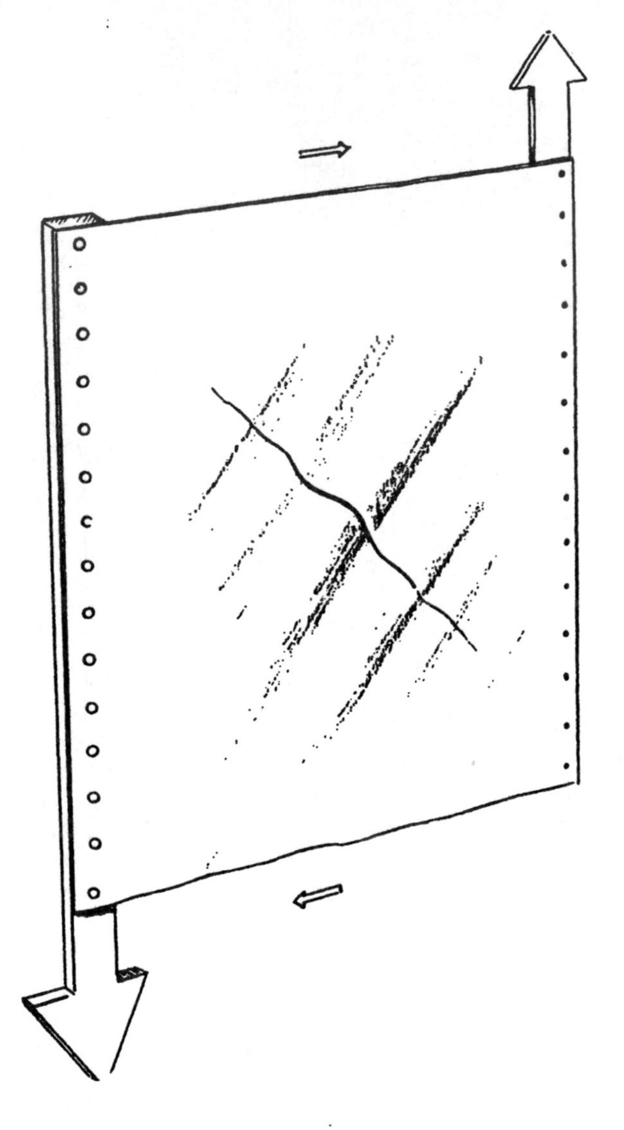

NOTE: Tension failure occurs after buckling is completed.
Tension failure is 90° to the direction of buckling.

***Figure A.12***

# BENDING LOAD - BUCKLING FAILURE

## HOLLOW DUCTILE METAL TUBE

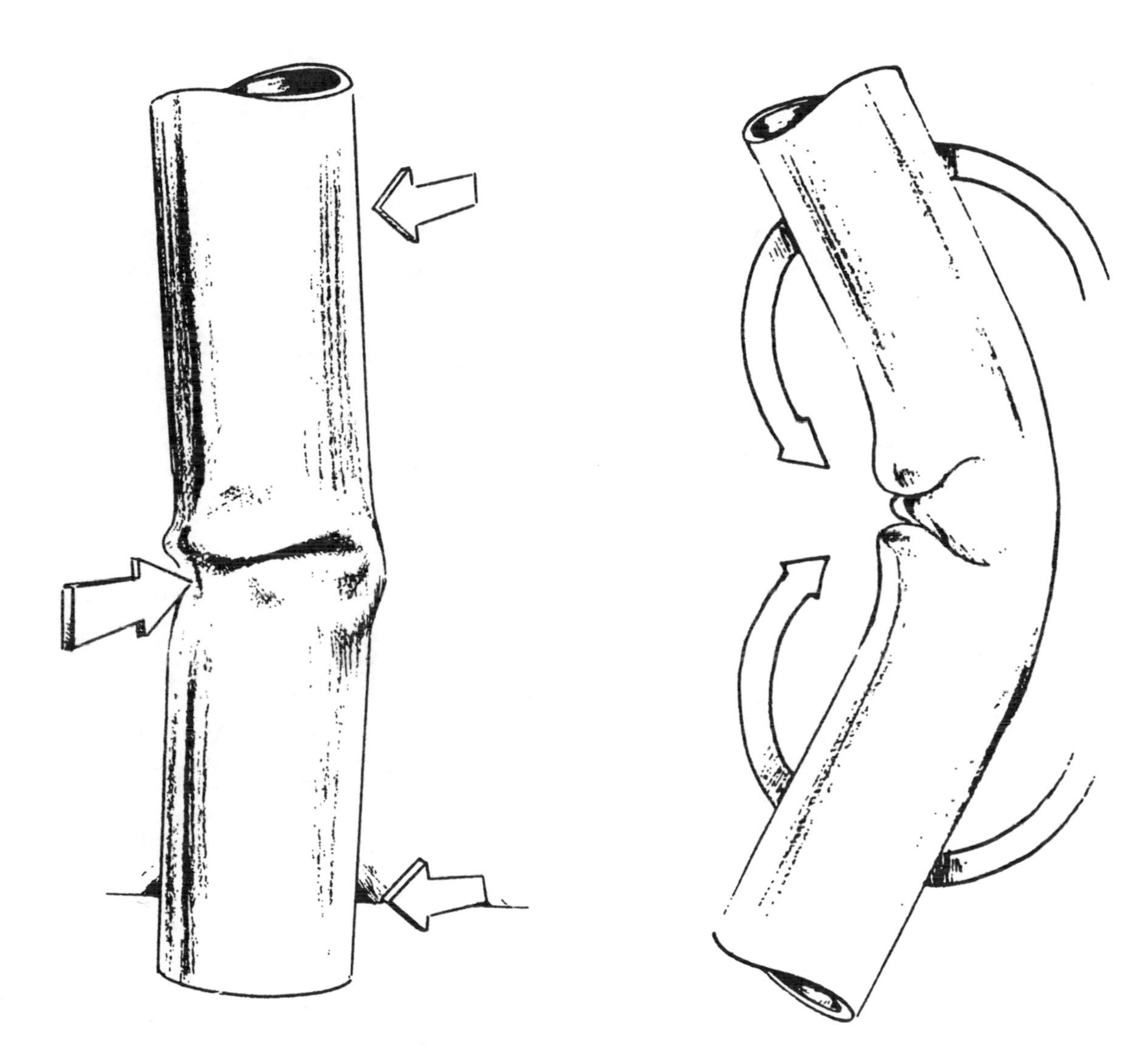

*Figure A.13*

# BENDING LOAD - TENSION FAILURE

## SOLID BRITTLE METAL SHAFT

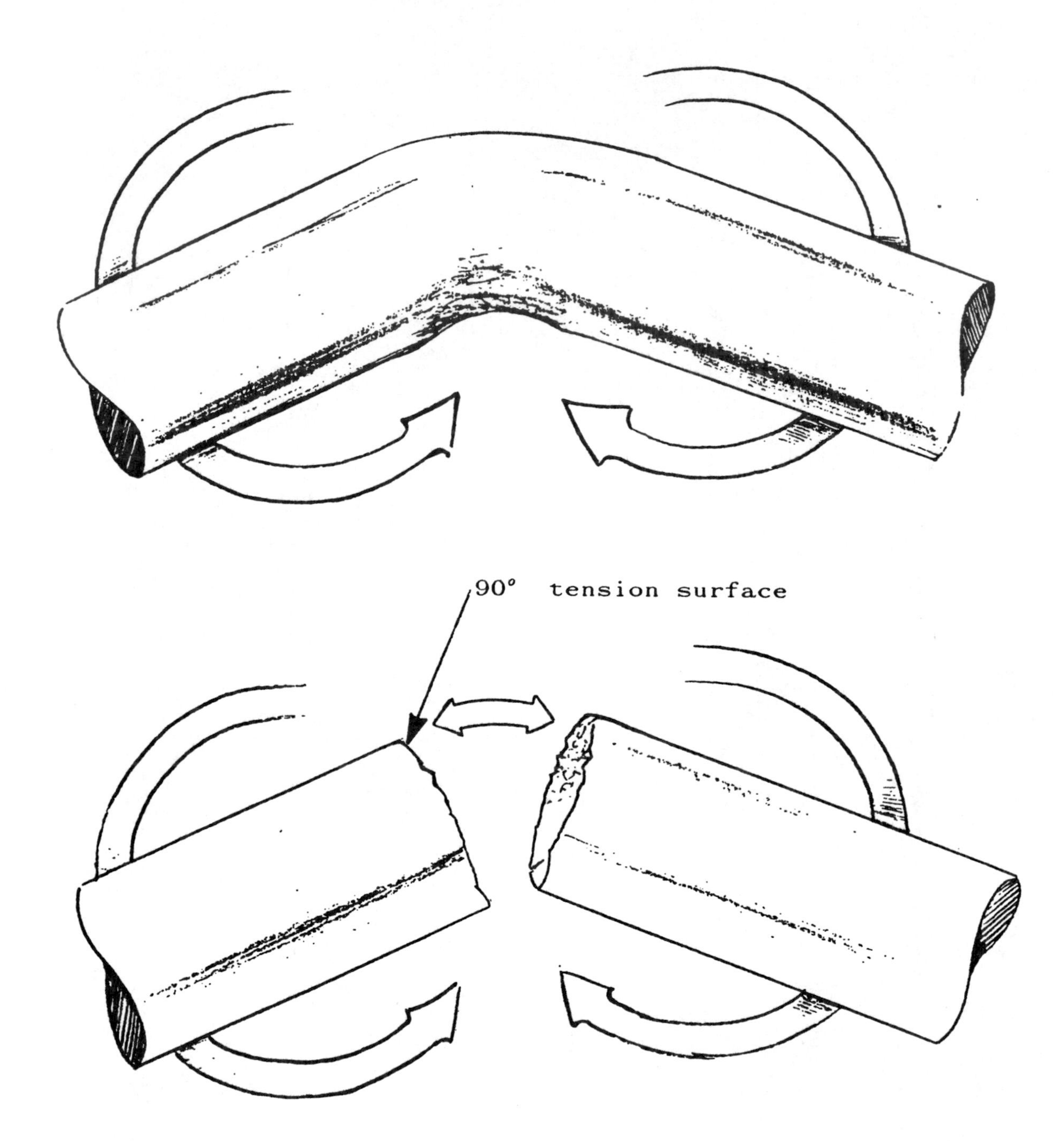

*Figure A.14*

# BENDING LOAD - SHEAR FAILURE

## SOLID DUCTILE METAL SHAFT

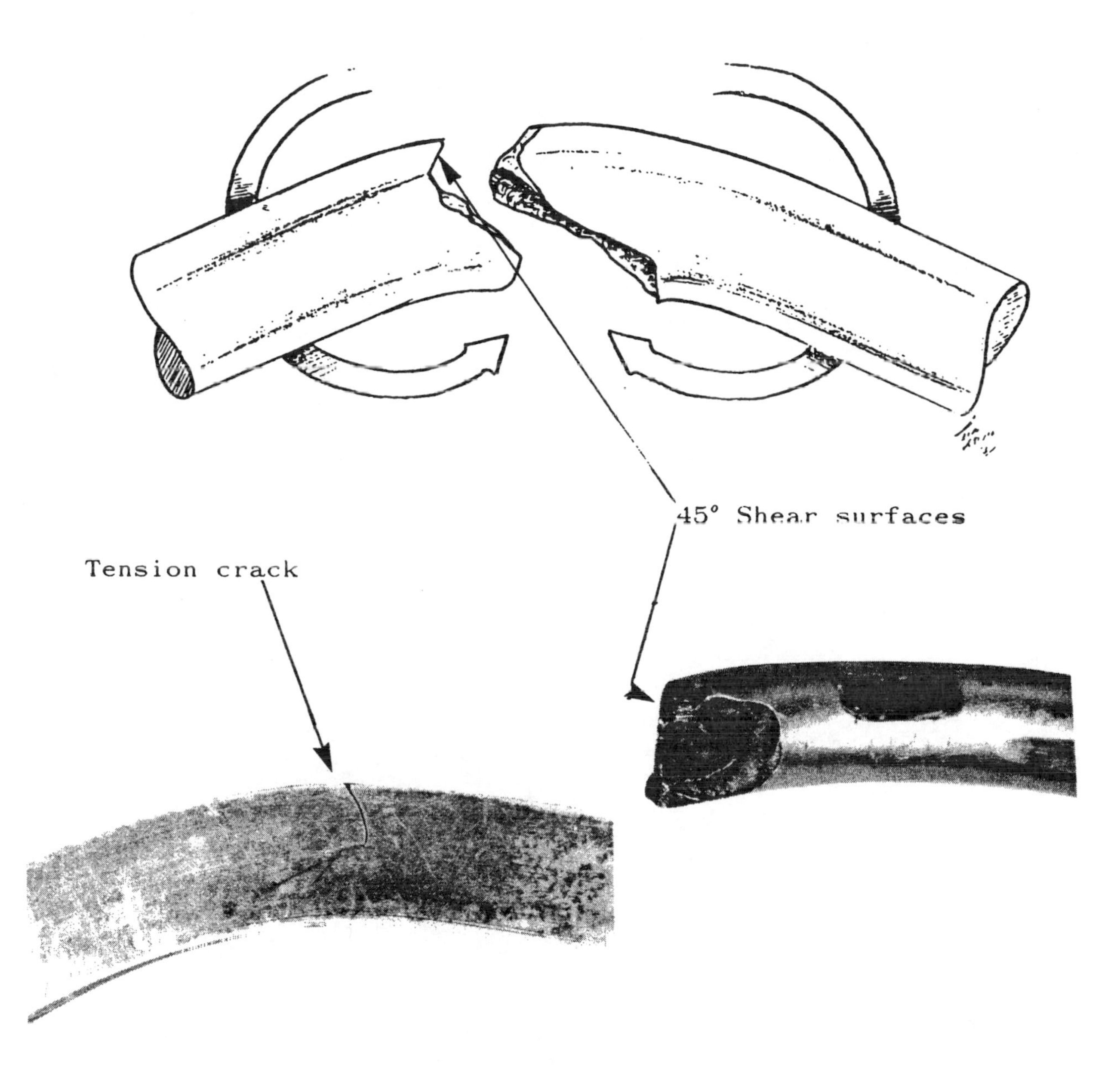

*Figure A.15*

# TORSION LOAD - SHEAR FAILURE

## SOLID DUCTILE METAL SHAFT

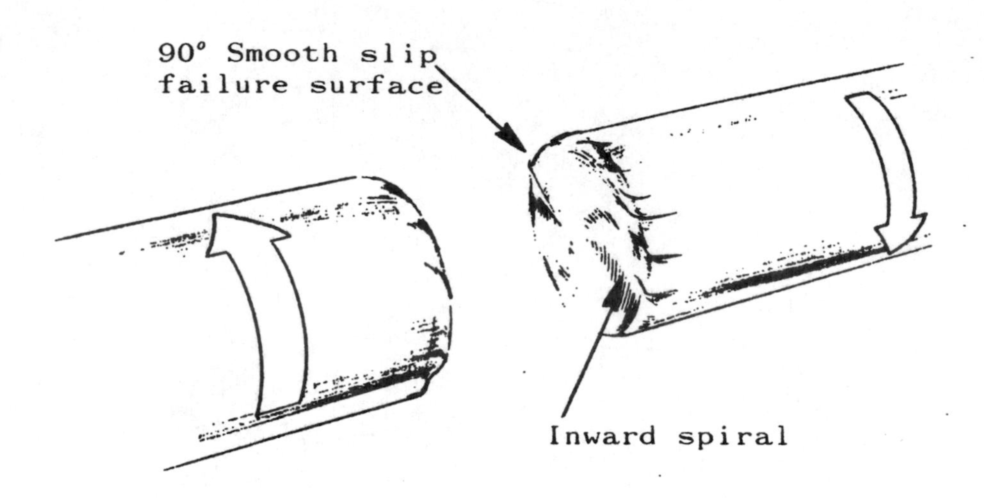


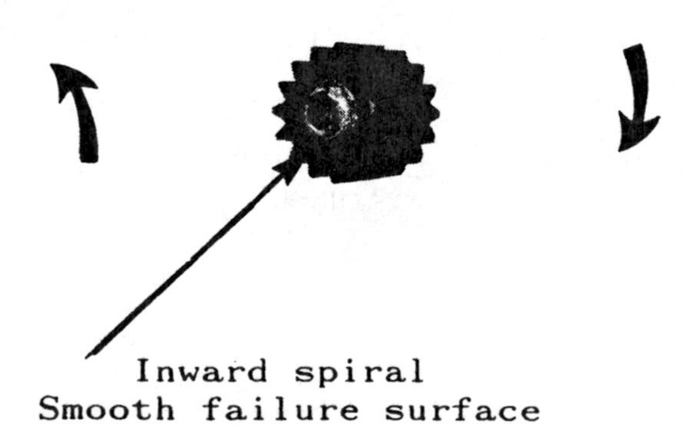


***Figure A.16***

# TORSION LOAD - COMPRESSION FAILURE

## HOLLOW DUCTILE METAL SHAFT

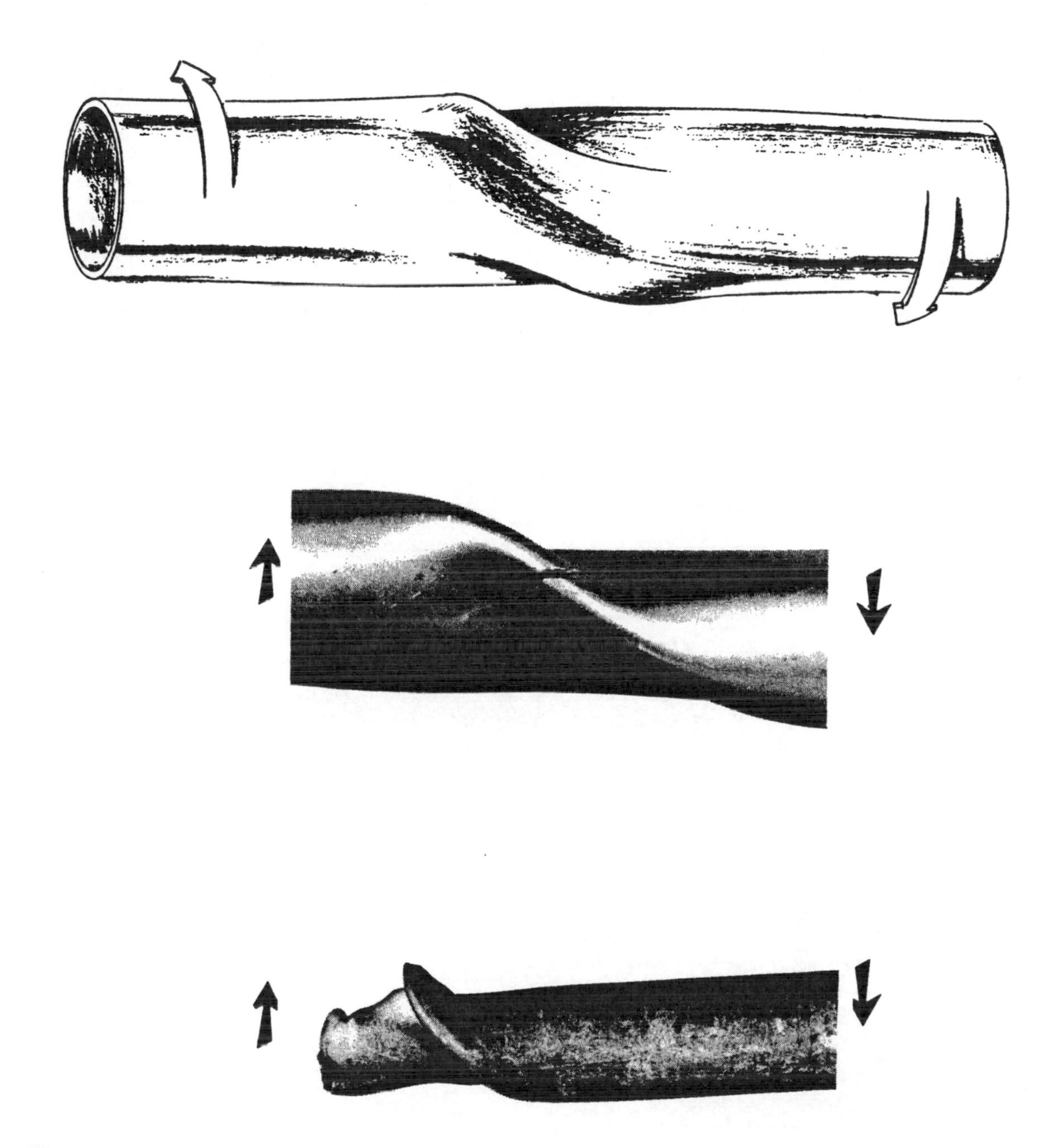

*Figure A.17*

# TORSION LOAD - TENSION FAILURE

## SOLID BRITTLE METAL SHAFT

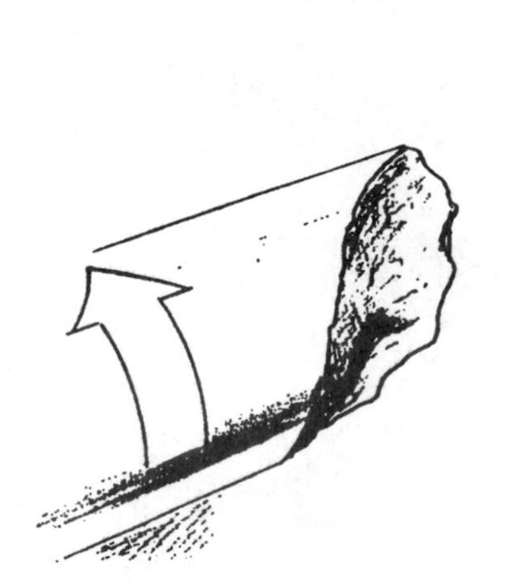
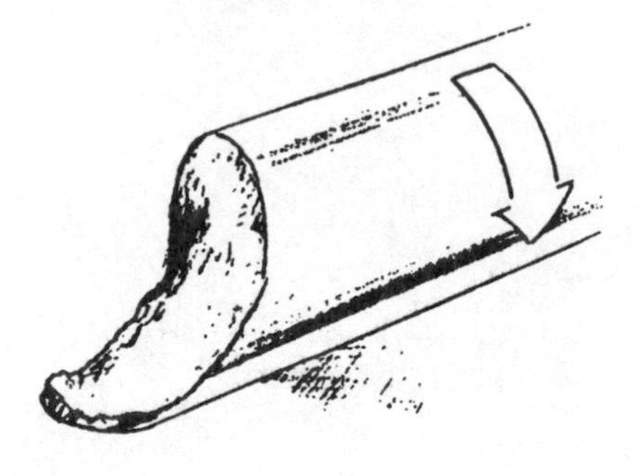

45° Helical rough tension failure surface

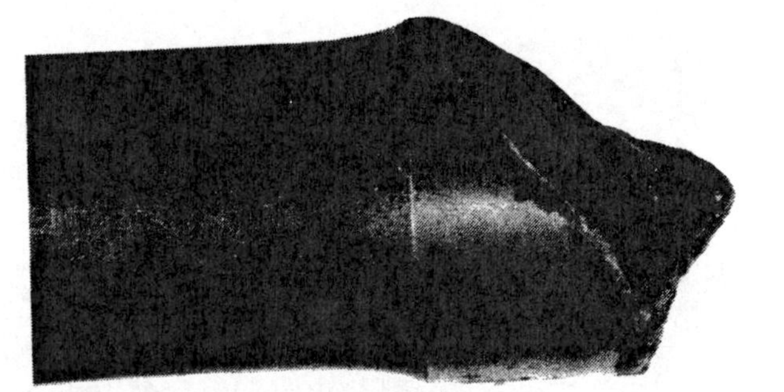

45°

Star Pattern variation

***Figure A.18***

# TORSION LOAD - TENSION FAILURE

## HOLLOW BRITTLE METAL SHAFT (rare in A/C)

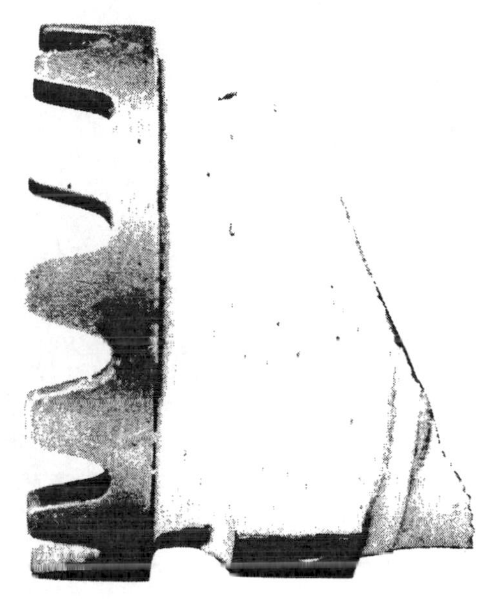

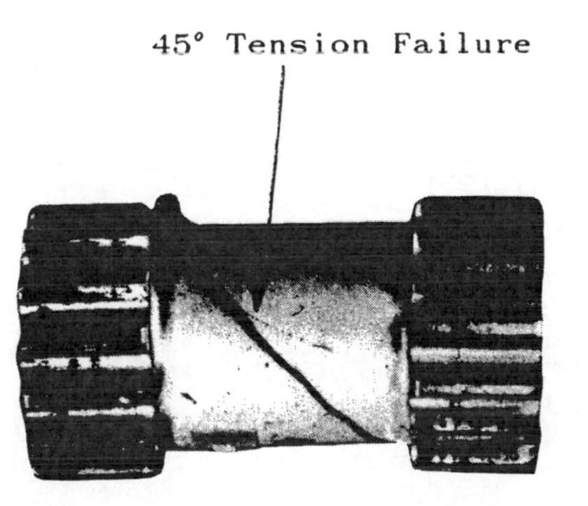


*Figure A.19*

# FATIGUE FAILURE - ONE WAY BENDING

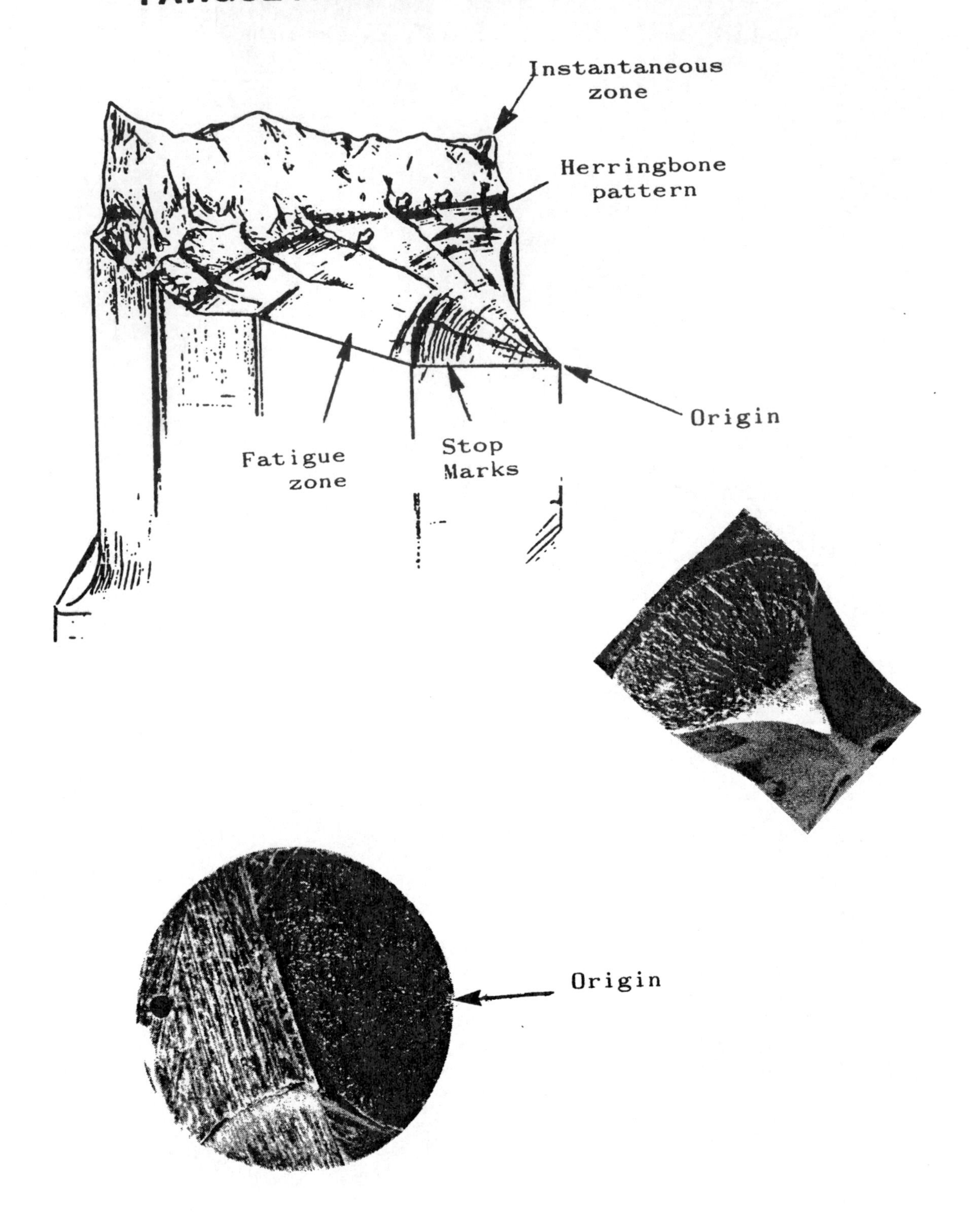


*Figure A.20*

# FATIGUE FAILURE - ONE WAY BENDING

## FAILURES STARTING FROM STRESS CONCENTRATIONS

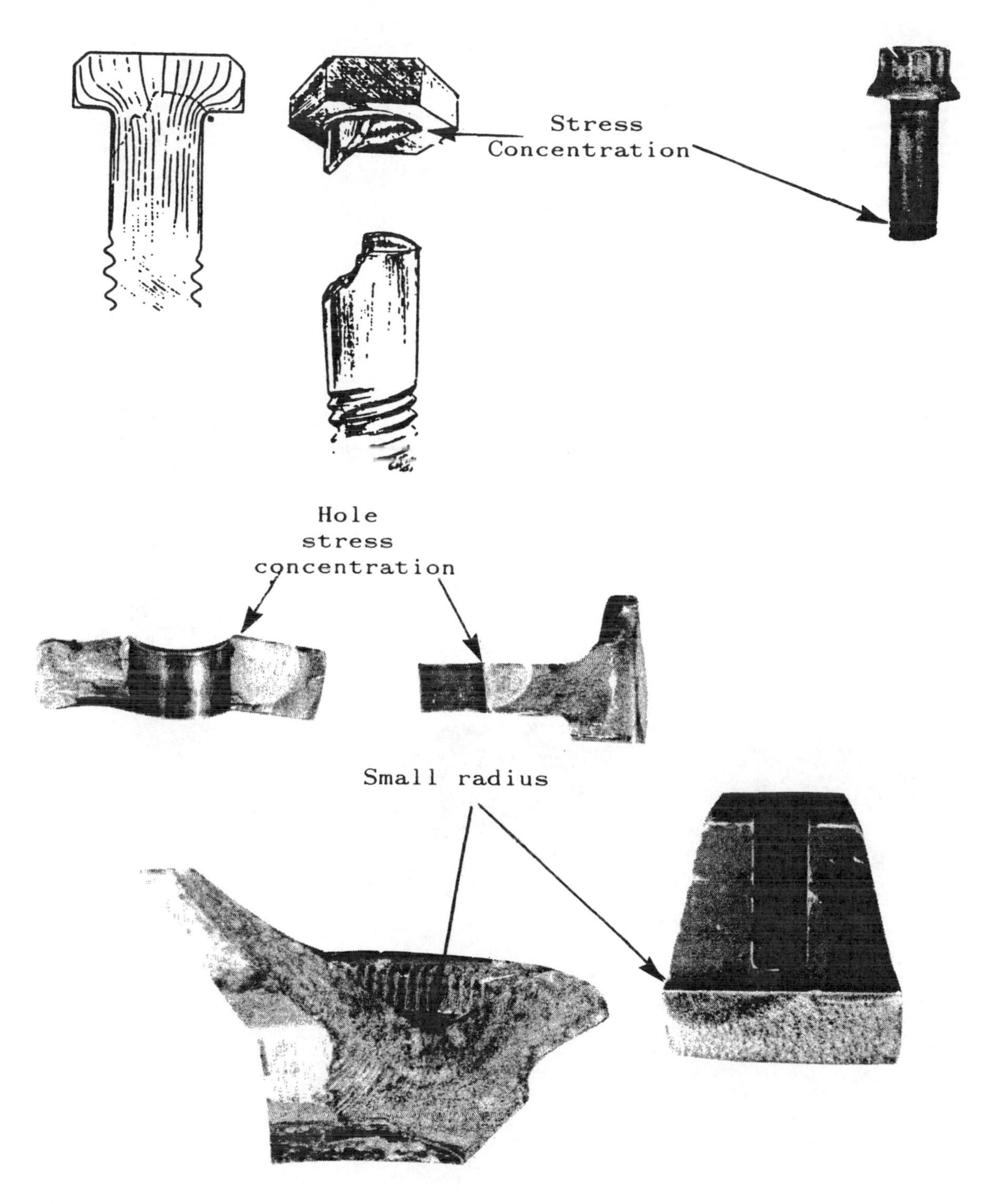

*Figure A.21*

# FATIGUE FAILURE - ONE WAY BENDING

(con'd)

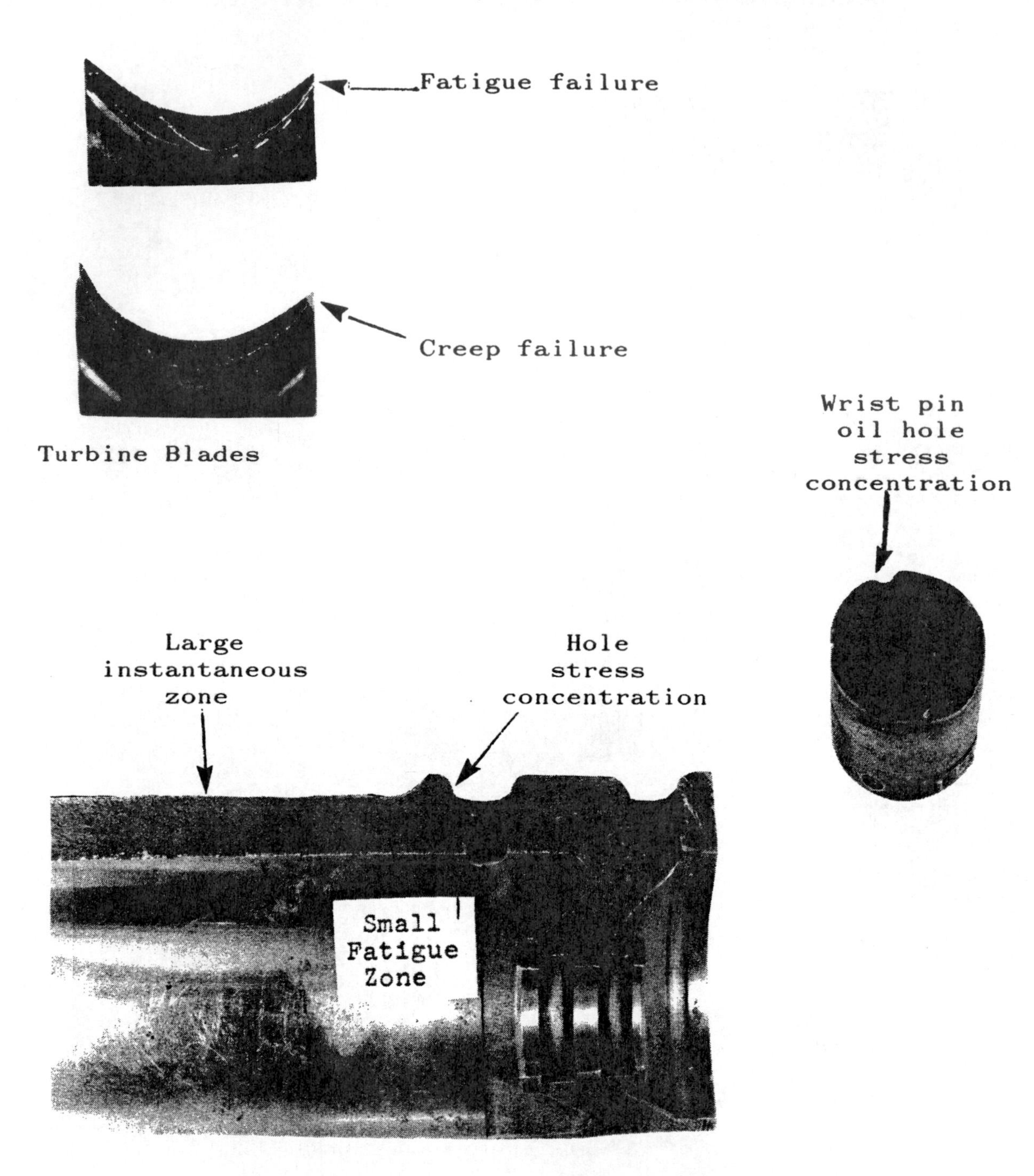


Hydraulic Cylinder

*Figure A.22*

# FATIGUE FAILURE - TWO WAY BENDING

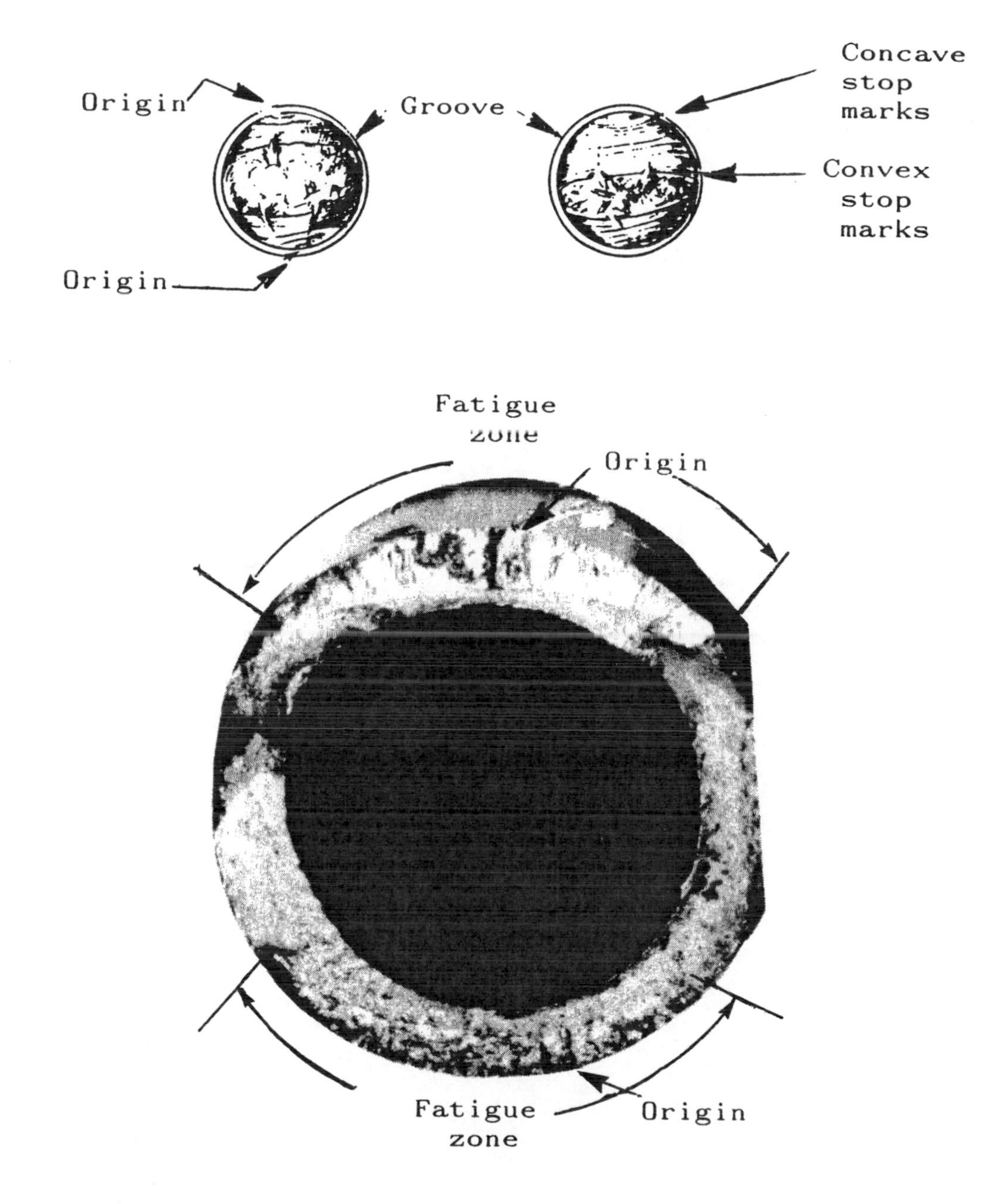


***Figure A.23***

# FATIGUE FAILURE - TORSION LOADING

## WITH NO STRESS CONCENTRATION PRESENT, FAILURE IS ON A PLANE 45° TO AXIS

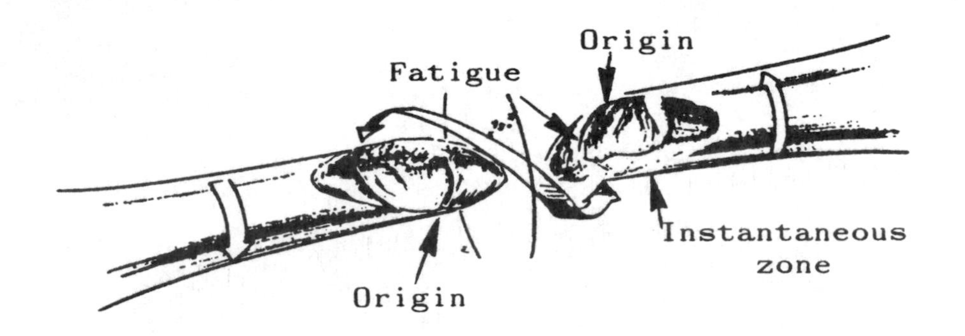


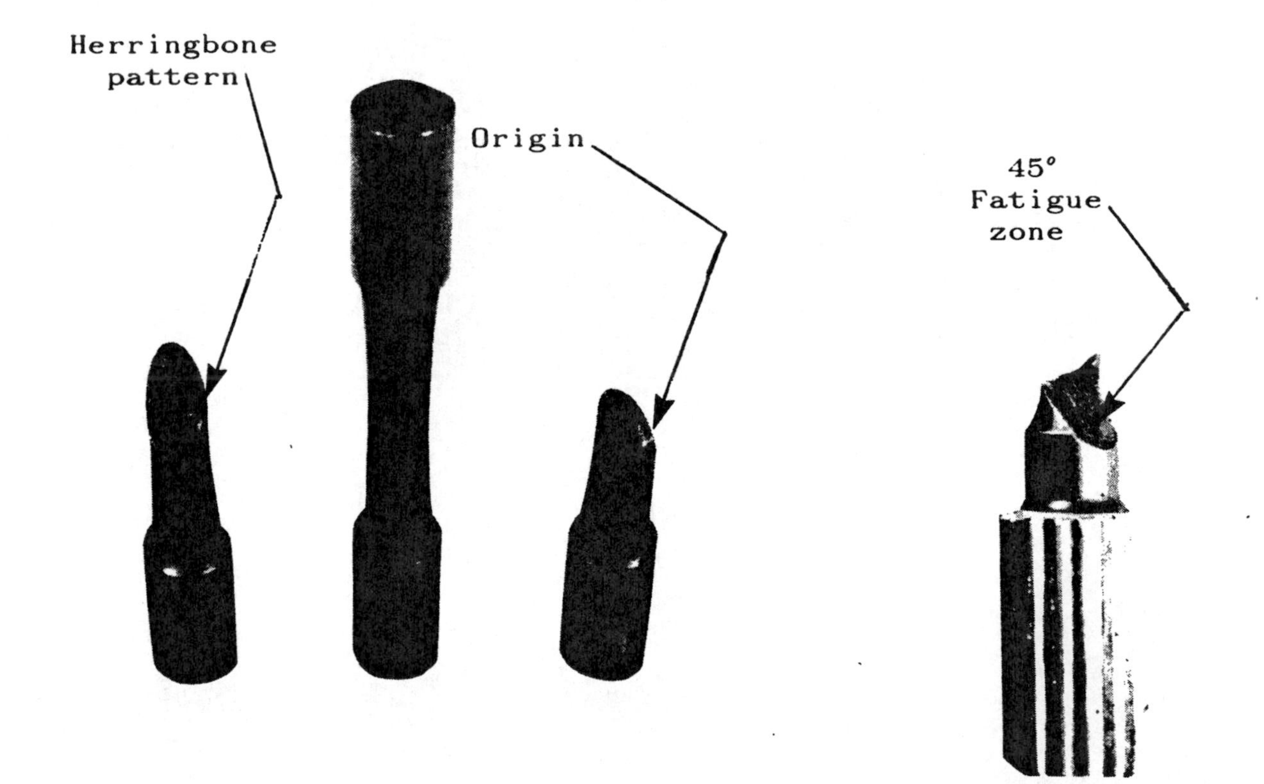

*Figure A.24*

# FATIGUE FAILURE - TORSION LOADING

## WITH A GROOVE STRESS CONCENTRATION PRESENT, FAILURE IS ON A PLANE 90° TO AXIS

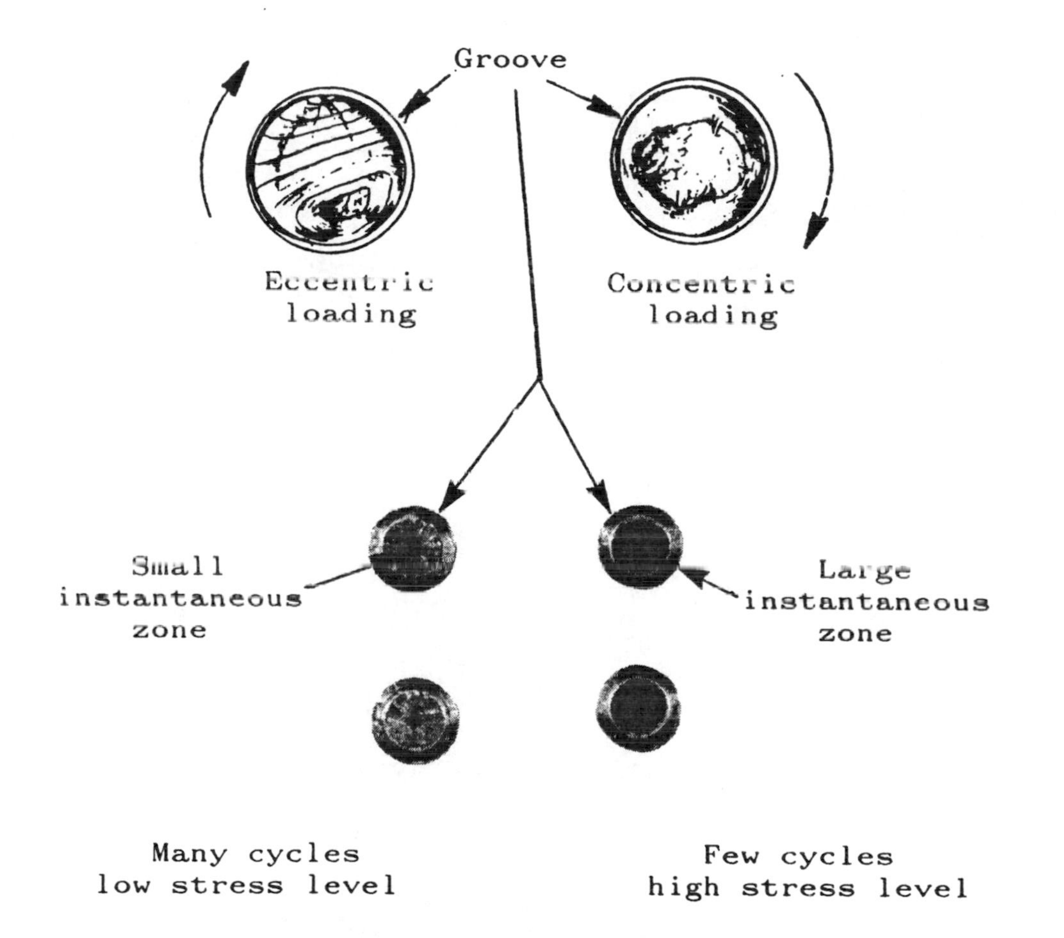


***Figure A.25***

# INTERGRANULAR CORROSION

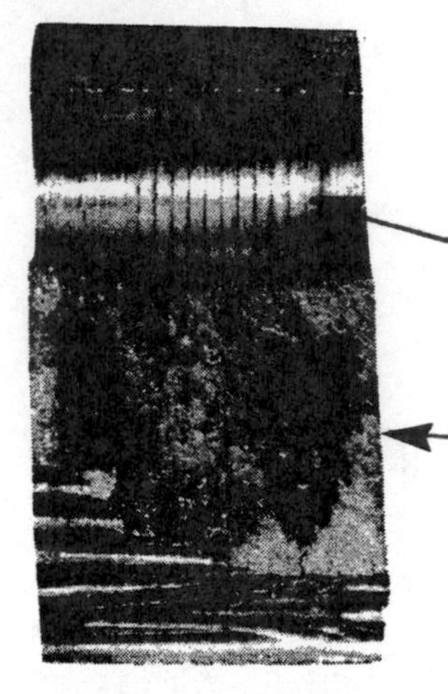

Corrosion started
at machined grooves

Corroded area

Intergranular Corrosion
started from surface
corrosion in hole

Corroded area

***Figure A.26***

# EXFOLIATION CORROSION

## (specialized intergranular)

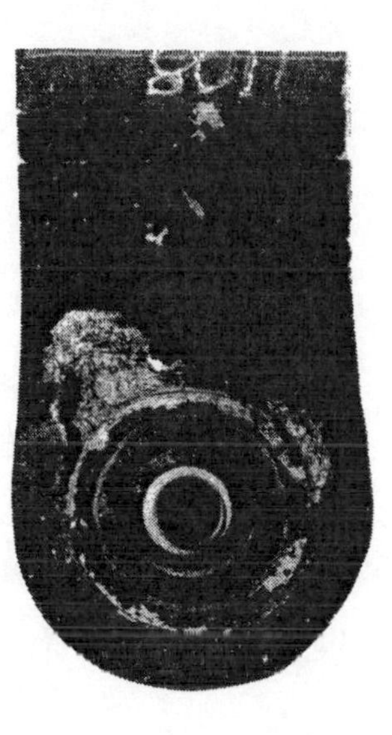

Raised "blistered" surface

Corrosion advances parallel to surface

***Figure A.27***

# CORROSION FATIGUE

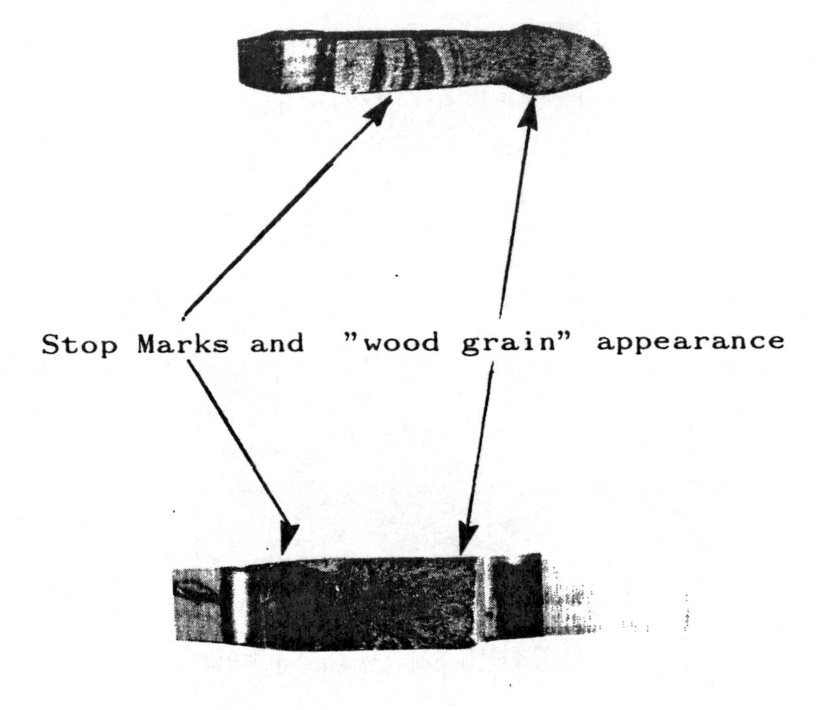


***Figure A.28***

# REFERENCES


1 AIAA, *22nd Structures, Structural Dynamics and Materials Conference*, American Institute of Aerodynamics and Astronautics, NY, 1981

2 Anonymous, *Corrosion*, Lockheed Field Service Digest 49, Lockheed California Co., Los Angeles, 1965

3 Anonymous, *Fatigue and Stress Corrosion*, Lockheed California Co., Los Angeles, 1968

4 Anonymous, *Mechanical Failures of Metals in Service*, NBS Circular 550, US Department of Commerce, National Bureau of Standards, Washington DC, 1954

5 Anonymous, *Nondestructive Testing Systems*, Magnaflux Corp., Chicago, 1968

6 Bruch, C.D., *Strength of Materials for Technology*, John Wiley & Sons, NY 1976

7 Bruhn, E.F., *Analysis and Design of Flight Vehicle Structures*, Tri-State Offset Co., Cincinnati, 1965

8 Craig, H.L.Jr. et al, *Corrosion-Fatigue Technology*, Pub 642 American Society for Testing Materials, Baltimore, 1976

9 Clauser, H.R., *Industrial and Engineering Materials*, McGraw Hill, NY 1975

10 Cugini, J.D., Kevlar, Aviation Mechanics J, June 1986

11 Fong, J.T.(Ed), *Fatigue Mechanisms*, STP 675, American Society for Testing Materials, Baltimore, 1979

12 Gray, A.G.(Ed), *Nondestructive Testing for Management*, American Society for Metals, Menlo Park, OH 1963

13 Grover, H.J., *Fatigue of Aircraft Structures*, NAVAIR 01-1A-13, Naval Air Systems Command, Dept. of the Navy, Washington DC, 1960

14 Guy, A.G. and J.J. Hren, *Elements of Physical Metallurgy*, 3rd Ed., Addison-Wesley Pub. Co., Reading, MA, 1973

15 Holt D.J.(Ed), *Arall*, Aerospace Engineering, May 1985

16 Hurt, H.H. Jr., *Aerodynamics for Naval Aviators*, NAVWEPS 00-80T-80, US Govt. Printing Office, 1959

## *References*

17 ______, *Fundamentals of Helicopter Structures*, Univ. of Southern California, Los Angeles, 1967

18 Jones, R.M., *Mechanics of Composite Materials*, Scriata Book Co., Washington DC, 1975

19 Kleivan, A. and A. Roed, *Flight Safety of Aircraft Structures and Systems*, Royal Institute of Technology, Stockholm, Sweden, undated

20 Lankford, J. et al (Eds), *Fatigue Mechanisms*, STP 811, American Society for Testing Materials, Baltimore, 1983

21 Madayag, A.F.(Ed), *Metal Fatigue Theory and Design*, John Wiley Sons, NY, 1969

22 Manson, S.S.(Ed), *Metal Fatigue Damage*, STP 811, American Society for Testing Materials, Philadelphia, 1971

23 Masters, J.E., *Basic Failure Modes of Continuous Fiber Composites*, Vol I, Engr. Matl. Handbook, ASM, Metals Park, OH, 1988

24 Osgood, C.C., *A Basic Course in Fracture Mechanics*, Machine Design, 1971

25 Perry, D.J., *Aircraft Structures*, McGraw-Hill, NY, 1950

26 Reinhart, T.J. and L.Cremets, *Introduction to Composites*, Vol I, Engr. Matl. Handbook, ASM, Metals Park, OH, 1988

27 Seely, F.B., *Resistance of Materials*, John Wiley Sons, NY, 1935

28 Shanley, F.R., *Strength of Materials*, McGraw-Hill, NY, 1957

29 ______, *Weight-Strength Analysis of Aircraft Structures*, Dover Publications, NY, 1951

30 Smoot, R.C. et al, *Chemistry, A Modern Course*, Merrill Publishing Co., Columbus, OH, 1971

31 Stone, F.R. Jr., *Metal Fatigue and Its Recognition*, Eng. Div. Bulletin No. 63-1, Civil Aeronautics Board, Bureau of Safety, Washington DC, 1963

32 Teichmann, F.K., *Fundamentals of Aircraft Structural Analysis*, Hayden Book Co., NY, 1968

# INDEX

## Index